SIMPLY

PHYSICS

Produced for DK by
cobalt id
www.cobaltid.co.uk

Design Director Paul Reid
Designers Clare Joyce, Mik Gates, Mark Lloyd
Editor Marek Walisiewicz
Creative Technical Support Darren Bland

Senior Editor Peter Frances
Editor Abigail Ellis
Senior Designer Jessica Tapolcai
Managing Editor Angeles Gavira Guerrero
Managing Art Editor Michael Duffy
Production Editor Andy Hilliard
Production Controller Rebecca Parton
Jacket Design Development Manager
Sophia M.T.T
Project Jacket Designer Juhi Sheth
Publishing Director Georgina Dee
Art Director Maxine Pedliham
Managing Director Liz Gough
Design Director Phil Ormerod

First published in Great Britain in 2025 by
Dorling Kindersley Limited
20 Vauxhall Bridge Road
London SW1V 2SA

The authorised representative in the EEA is
Dorling Kindersley Verlag GmbH. Arnulfstr. 124,
80636 Munich, Germany

A CIP catalogue record for this book
is available from the British Library.
ISBN: 978-0-2417-2275-6

Printed and bound in China

www.dk.com

CONSULTANTS AND CONTRIBUTORS

Jack Challoner is the author of more than 50 books on science and technology. Before becoming a writer, he worked at London's Science Museum. He studied physics and trained as a science and maths teacher.

Hilary Lamb is an award-winning science journalist, editor, and author. She has worked as an author and consultant on more than a dozen DK books, including *The Physics Book* and *Simply Quantum Physics*, and edits the literary magazine *Tamarind*.

David Sang is an experienced teacher of physics and has contributed to over 100 textbooks, as well as teaching guides and websites.

Giles Sparrow is an author and journalist specializing in astronomy and space exploration. He has written dozens of books, and is a Fellow of the Royal Astronomical Society.

CONTENTS

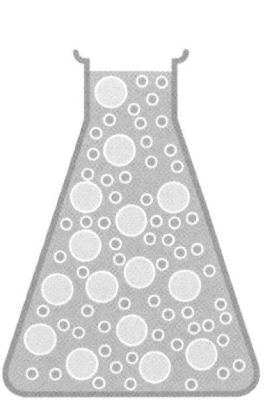
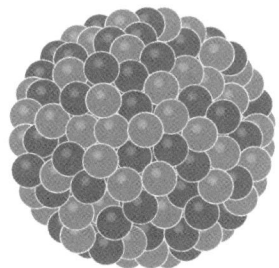
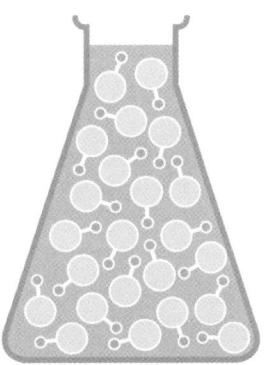

FORCES AND MOTION

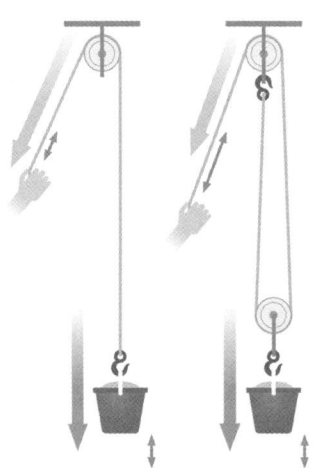

ENERGY

ELECTRICITY AND MAGNETISM

SOUND
AND LIGHT

QUANTUM
PHYSICS

RELATIVITY

ASTROPHYSICS
AND **COSMOLOGY**

WHAT IS PHYSICS?

Physics is a way of understanding some of the most fundamental phenomena in the world around us. Its subjects are matter and energy, here on Earth and throughout the Universe. The explanatory ideas that physicists have developed over the centuries are based on what they see in the natural world. Sometimes they simply observe natural phenomena; sometimes experiments are set up to answer questions that arise from observing nature. Experiments may be as simple as bringing two magnets together or as complex as the giant particle colliders used to investigate the composition of matter at the smallest scale. Experimental evidence is the ultimate test of any physics theory.

In physics, new discoveries can open up entirely new fields of research. The idea that matter is made up of particles enabled scientists to develop explanations of the behaviour of solids, liquids, and gases. Understanding the detailed structure of atoms showed how atoms can bind together to make the myriad molecules studied by chemists and biologists. Understanding how light interacts with atoms allowed astronomers to discover the composition of stars. The more we know, the more there is to find out and the more tools we have to make discoveries.

While physicists are excited by the theories of physics, their application has dramatically changed human lives through technology. Our needs for shelter, transport, communication, healthcare, even entertainment have all been transformed by the application of physicists' understanding of how things work.

MATTE

R

Everything around us is made of the stuff we call matter. At first sight, it may seem that water, air, and rocks are very different, but at the fundamental level they are all made of the same stuff – atoms, which are in turn made up of even smaller components. Studies of starlight show that everything we can see in space – planets, galaxies, and so on – is also made of the same matter as things on Earth. The tiny particles that make up matter move and interact with one another according to a few basic laws, which people have come to understand over the last four centuries. To a physicist, the Universe is nothing but particles and their interactions.

STATES OF MOTION

Solids have a fixed shape and size; liquids flow but their volume remains constant; while gases spread out to fill the space available to them. These observations can be explained by thinking of matter as being made up of particles in constant, rapid motion. The particles attract their neighbours and tend to clump together to a degree that depends on how much energy they have. This way of thinking about particles is known as kinetic theory.

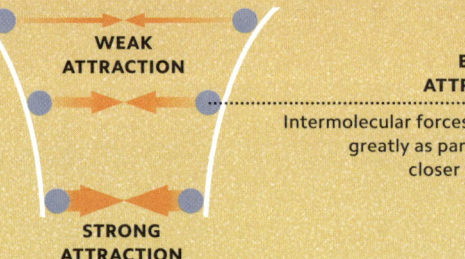

WEAK ATTRACTION

STRONG ATTRACTION

ELECTRIC ATTRACTION

Intermolecular forces increase greatly as particles get closer together.

Forces between particles

Intermolecular forces are far much weaker than the bonds that hold molecules together, but are key in determining state of matter.

SUBLIMATION

DEPOSITION

FREEZING

CRYSTALLINE SOLID

Solid

In a solid, particles vibrate but are held in position relative to their neighbours by strong intermolecular forces. In some solids, the particles are arranged in a repeating order, forming crystals.

Changing state

When matter is heated, its particles move faster. This may allow them to overcome the intermolecular forces so that the substance changes state from liquid to gas, for example. These changes are purely physical and do not alter the particles' chemical properties.

Gas

The particles of a gas move rapidly and in random directions, filling the volume available to them.

PRESSURIZED GAS

BOILING

CONDENSATION

At room temperature, atoms of helium gas move at speeds of more than 4,000kph (2,500mph).

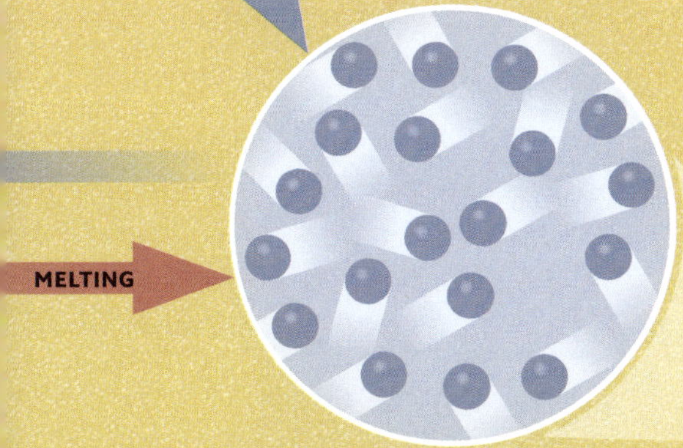

MELTING

Liquid

In a liquid at room temperature, particles move rapidly; gaps between particles let them move past their neighbours. This allows liquids to flow and assume the shapes of their containers.

LIQUID

GAS BEHAVIOUR

The invention of pumps, barometers, and thermometers in the 17th century allowed scientists to study the behaviour of gases. Experiments gave rise to a set of laws that set out the relationships between the volume of a gas, its temperature, and its pressure (the force with which it pushes on a unit area). These laws were established long before kinetic theory (see pp.10–11) was devised, and made the – erroneous – assumption that there are no forces between the molecules of a gas. However, in practice, the forces between gas particles are so small that the laws do predict the behaviour of gases very well at temperatures significantly above the boiling point.

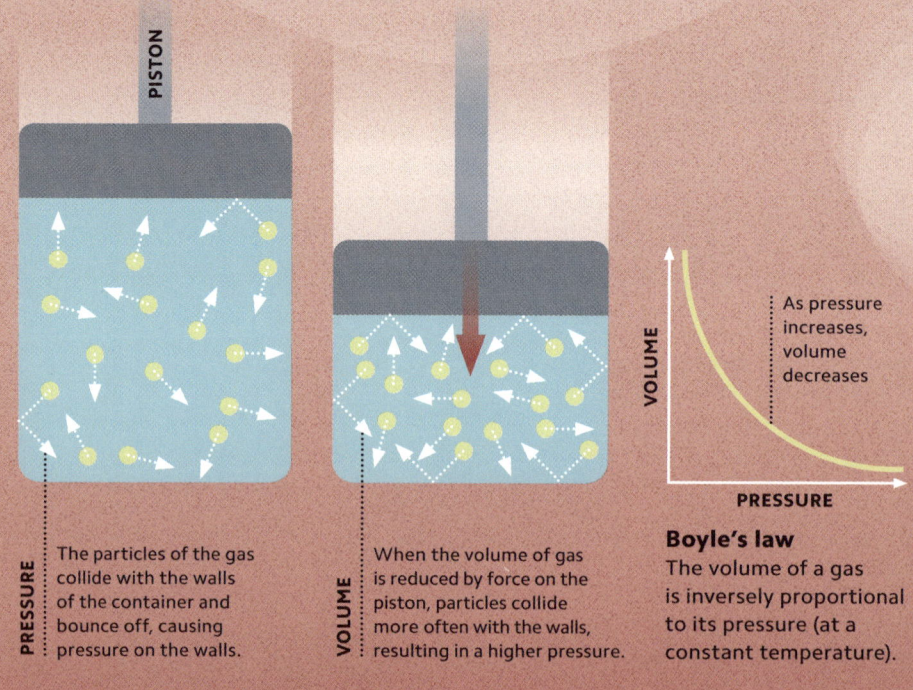

PISTON

PRESSURE

The particles of the gas collide with the walls of the container and bounce off, causing pressure on the walls.

VOLUME

When the volume of gas is reduced by force on the piston, particles collide more often with the walls, resulting in a higher pressure.

VOLUME

PRESSURE

As pressure increases, volume decreases

Boyle's law
The volume of a gas is inversely proportional to its pressure (at a constant temperature).

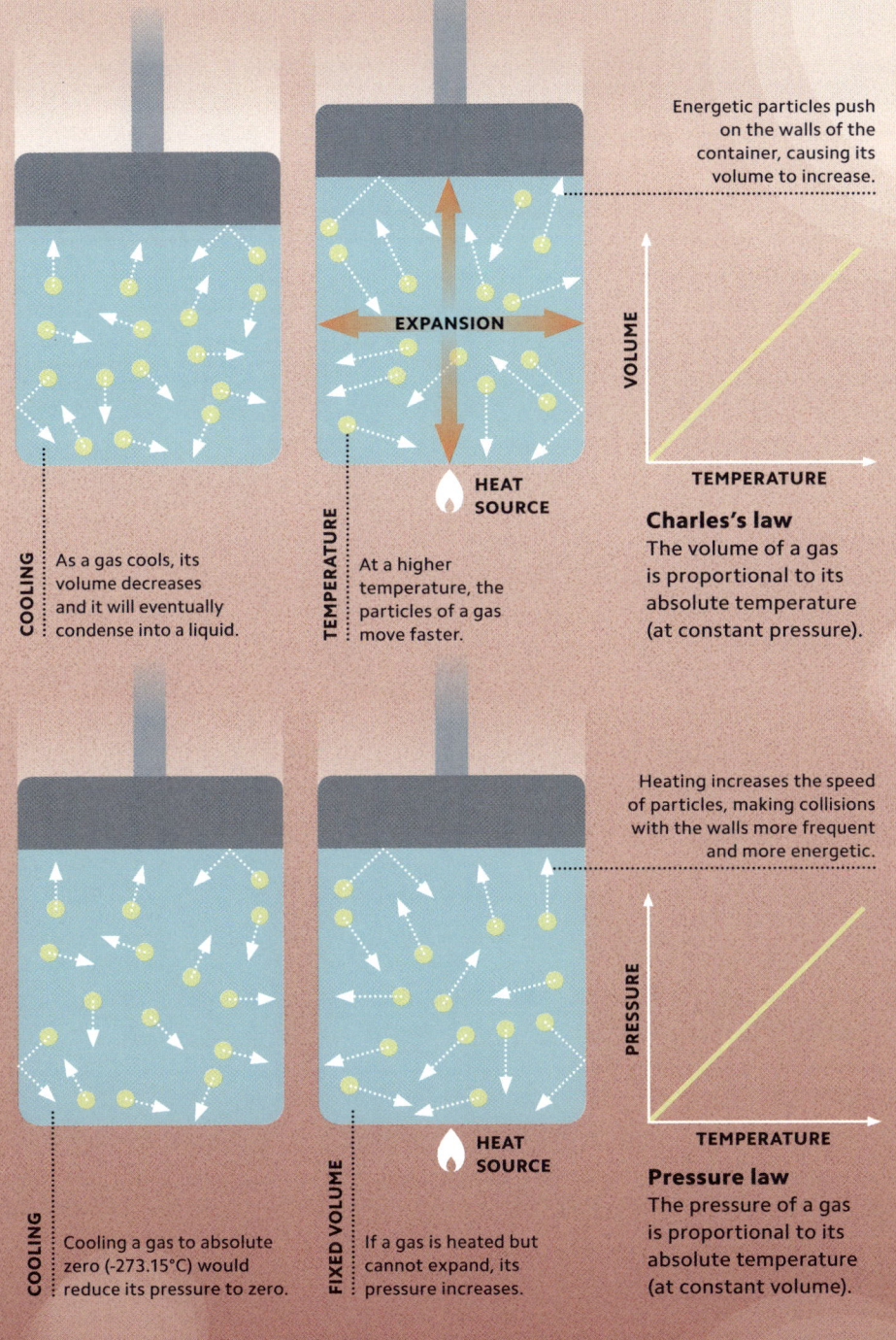

Energetic particles push on the walls of the container, causing its volume to increase.

EXPANSION

HEAT SOURCE

COOLING
As a gas cools, its volume decreases and it will eventually condense into a liquid.

TEMPERATURE
At a higher temperature, the particles of a gas move faster.

VOLUME

TEMPERATURE

Charles's law
The volume of a gas is proportional to its absolute temperature (at constant pressure).

Heating increases the speed of particles, making collisions with the walls more frequent and more energetic.

COOLING
Cooling a gas to absolute zero (-273.15°C) would reduce its pressure to zero.

FIXED VOLUME
If a gas is heated but cannot expand, its pressure increases.

HEAT SOURCE

PRESSURE

TEMPERATURE

Pressure law
The pressure of a gas is proportional to its absolute temperature (at constant volume).

INSIDE THE ATOM

All matter is made up of atoms – particles a fraction of a billionth of a metre across. Within each atom is a central nucleus composed of smaller particles called protons and neutrons. This is surrounded by a cloud of electrons. Each atom of an element has the same number of protons in its nucleus; different elements have different numbers of protons. This concept of the atom has allowed scientists to explain the differences between the elements, how elements combine to form compounds, and why some are radioactive (see pp.20–21).

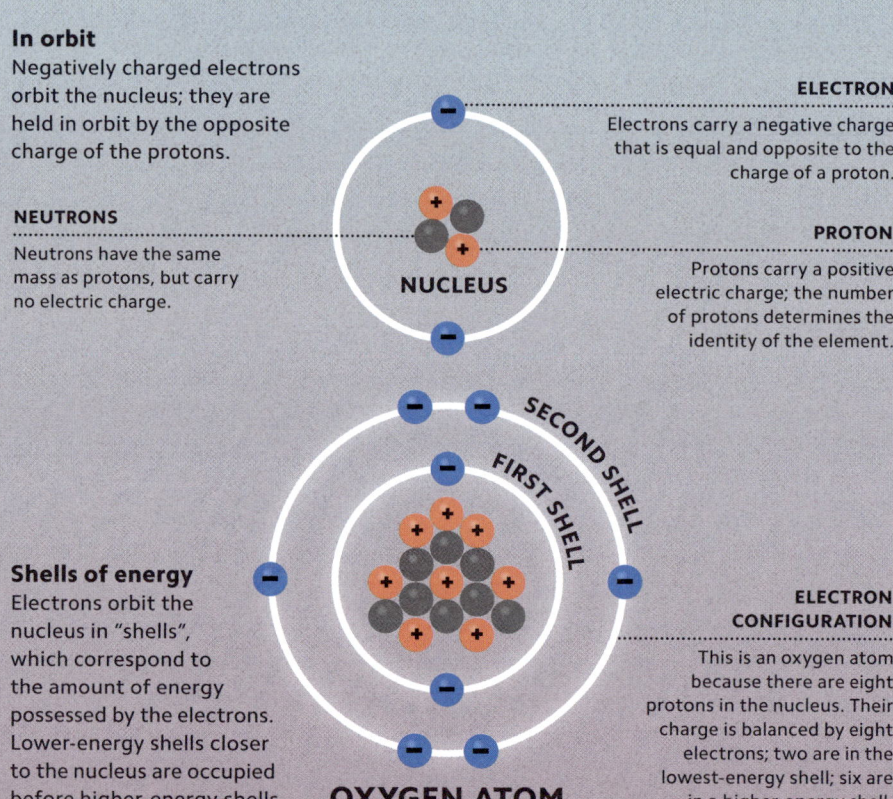

In orbit
Negatively charged electrons orbit the nucleus; they are held in orbit by the opposite charge of the protons.

NEUTRONS
Neutrons have the same mass as protons, but carry no electric charge.

ELECTRON
Electrons carry a negative charge that is equal and opposite to the charge of a proton.

PROTON
Protons carry a positive electric charge; the number of protons determines the identity of the element.

NUCLEUS

SECOND SHELL

FIRST SHELL

Shells of energy
Electrons orbit the nucleus in "shells", which correspond to the amount of energy possessed by the electrons. Lower-energy shells closer to the nucleus are occupied before higher-energy shells.

ELECTRON CONFIGURATION
This is an oxygen atom because there are eight protons in the nucleus. Their charge is balanced by eight electrons; two are in the lowest-energy shell; six are in a higher-energy shell.

OXYGEN ATOM

PUTTING THINGS TOGETHER

There are 94 naturally occurring elements – the fundamental substances that make up matter. Some matter, such as iron or diamond, is made of just one element, but most substances are made of several. Atoms of one element may be chemically bonded to other atoms to form compounds. Substances may also be mixed together, which does not involve making chemical bonds.

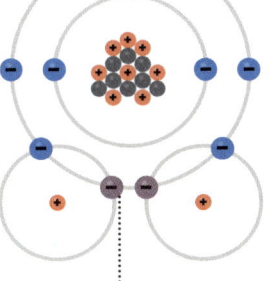

OXYGEN ATOM
The oxygen in a water molecule shares two electrons with the hydrogen atoms.

Making a bond
Many familiar substances are compounds. Water, for example, is made of hydrogen atoms bonded to oxygen atoms. The bond, known as a covalent bond, is formed by sharing electrons, and the resulting unit is called a molecule.

HYDROGEN ATOM

The two hydrogen atoms in a water molecule each share one electron with the oxygen atom.

WATER MOLECULE
Sharing electrons fills the outer electron shell of each atom, making the molecule stable.

Mixtures and compounds
Mixtures can be made up of varying proportions of different substances. Compounds always have the same ratio of constituent atoms.

Mixtures contain variable ratios of elements.

Elements in a compound are bound together in a fixed ratio.

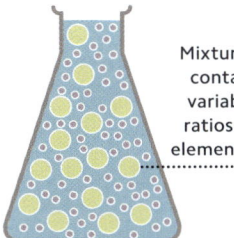

MIXTURE

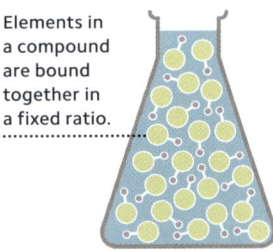

COMPOUND

ELECTRON TRANSFER

An atom contains the same number of protons as electrons. Their equal but opposite charges are balanced, so the atom has no overall electric charge. Some atoms, however, will readily give up one or more electrons from their outer shell. This loss of negative charge results in a positive ion. Others will readily accept electrons into their outer shell, so becoming a negative ion. Many compounds rely on ionic bonds between positive and negative ions.

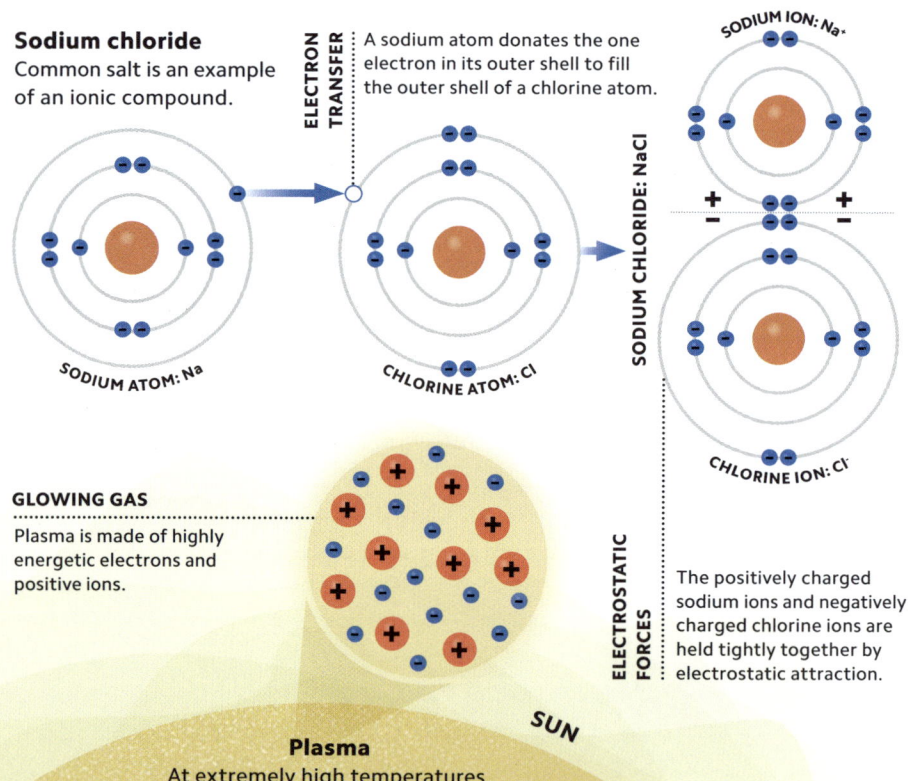

Sodium chloride
Common salt is an example of an ionic compound.

ELECTRON TRANSFER

A sodium atom donates the one electron in its outer shell to fill the outer shell of a chlorine atom.

SODIUM ATOM: Na

CHLORINE ATOM: Cl

SODIUM ION: Na⁺

CHLORINE ION: Cl⁻

SODIUM CHLORIDE: NaCl

ELECTROSTATIC FORCES

The positively charged sodium ions and negatively charged chlorine ions are held tightly together by electrostatic attraction.

GLOWING GAS

Plasma is made of highly energetic electrons and positive ions.

SUN

Plasma
At extremely high temperatures, electrons can be torn away from the atoms, forming an ionized gas called plasma. This is formed in stars and comprises 99 per cent of the visible universe.

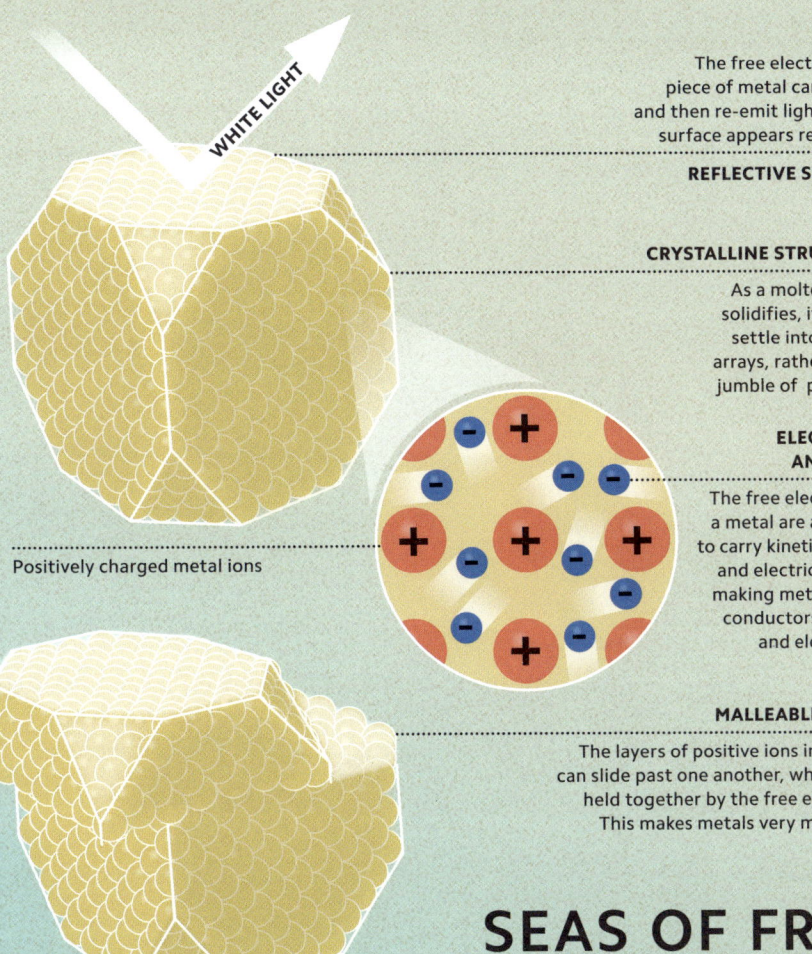

The free electrons in a piece of metal can absorb and then re-emit light, so the surface appears reflective.

REFLECTIVE SURFACE

CRYSTALLINE STRUCTURE

As a molten metal solidifies, its atoms settle into regular arrays, rather than a jumble of particles.

ELECTRONS AND IONS

The free electrons in a metal are available to carry kinetic energy and electric charge, making metals good conductors of heat and electricity.

WHITE LIGHT

Positively charged metal ions

MALLEABLE METAL

The layers of positive ions in a metal can slide past one another, while being held together by the free electrons. This makes metals very malleable.

SEAS OF FREE ELECTRONS

Three quarters of all the elements are metals – substances distinguished by the way their atoms bind together. The outer electrons of a metal atom are only loosely bound to their parent atoms, and become detached, forming a "sea" of electrons that can flow around a regular lattice of positively charged metal ions. This arrangement determines the physical properties of metals.

BEND ME, SHAPE ME

The materials used to make products are chosen for their physical properties, such as stiffness, toughness, and elasticity. These properties depend on the types of molecules that make up the solid, how they are arranged, and the forces between them.

Stress and strain

Stress is the force applied, per unit area, to a material; strain is the amount of deformation caused by stressing an object. The relationship between stress and strain determines how a material behaves and its suitability for different uses.

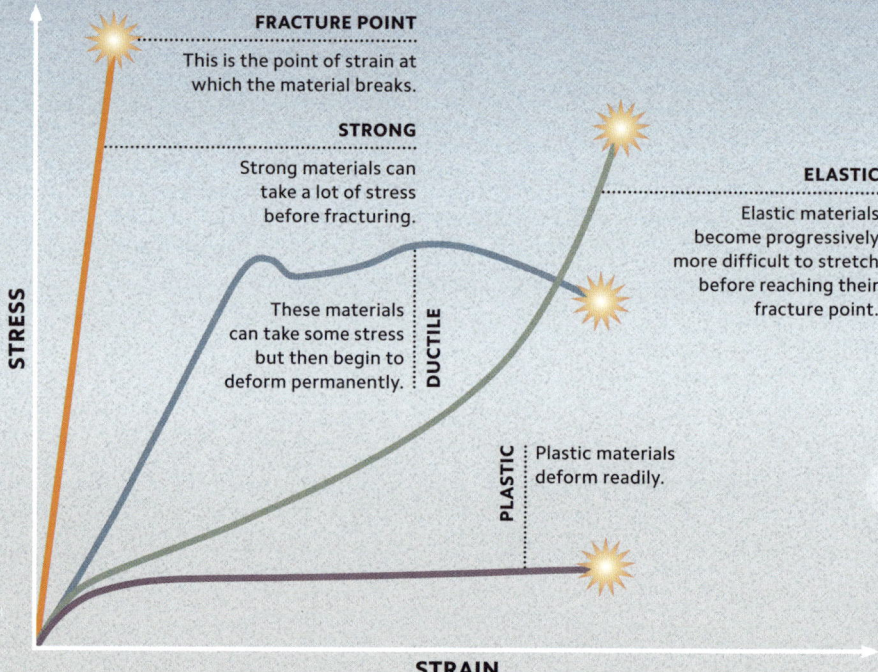

FRACTURE POINT
This is the point of strain at which the material breaks.

STRONG
Strong materials can take a lot of stress before fracturing.

ELASTIC
Elastic materials become progressively more difficult to stretch before reaching their fracture point.

These materials can take some stress but then begin to deform permanently.

DUCTILE

PLASTIC
Plastic materials deform readily.

STRESS

STRAIN

Material properties

Physicists use a range of words to describe the properties of solids. Most of these have more specific meanings compared with their everyday use.

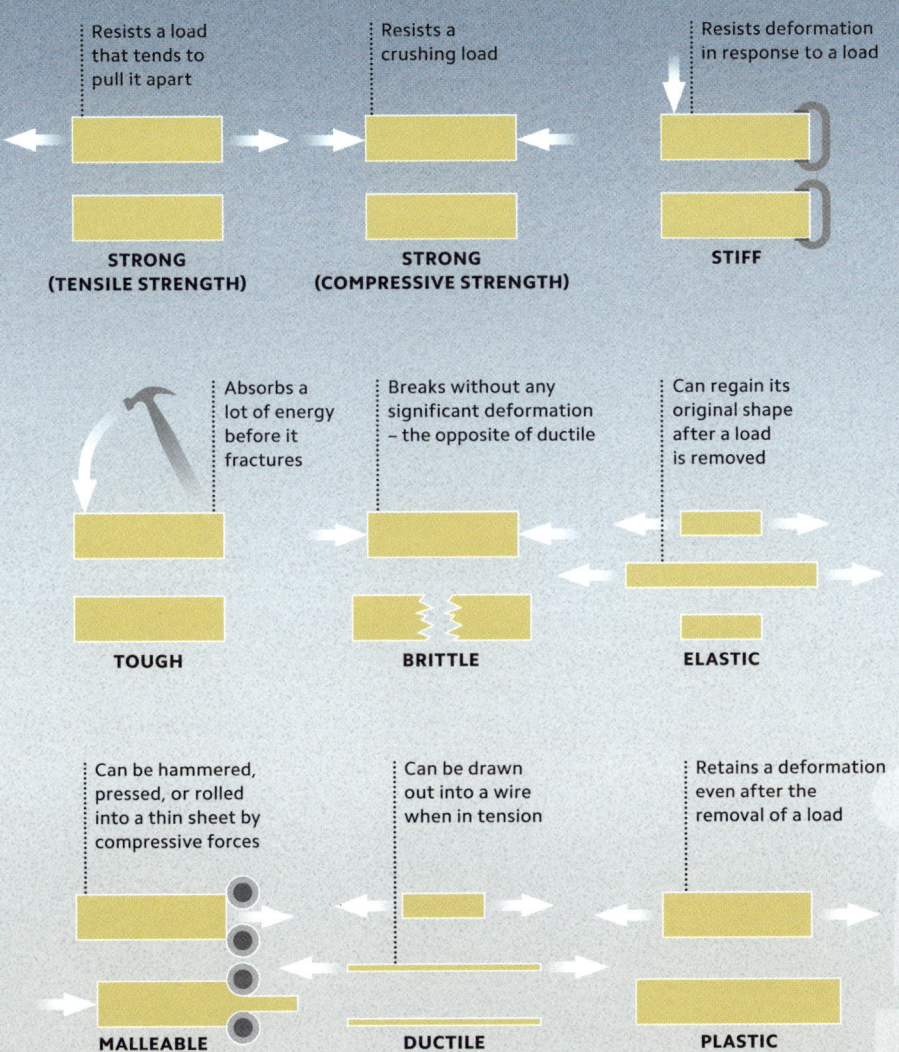

Resists a load that tends to pull it apart

STRONG (TENSILE STRENGTH)

Resists a crushing load

STRONG (COMPRESSIVE STRENGTH)

Resists deformation in response to a load

STIFF

Absorbs a lot of energy before it fractures

TOUGH

Breaks without any significant deformation – the opposite of ductile

BRITTLE

Can regain its original shape after a load is removed

ELASTIC

Can be hammered, pressed, or rolled into a thin sheet by compressive forces

MALLEABLE

Can be drawn out into a wire when in tension

DUCTILE

Retains a deformation even after the removal of a load

PLASTIC

Radioactive isotopes are naturally present in rocks and in the air.

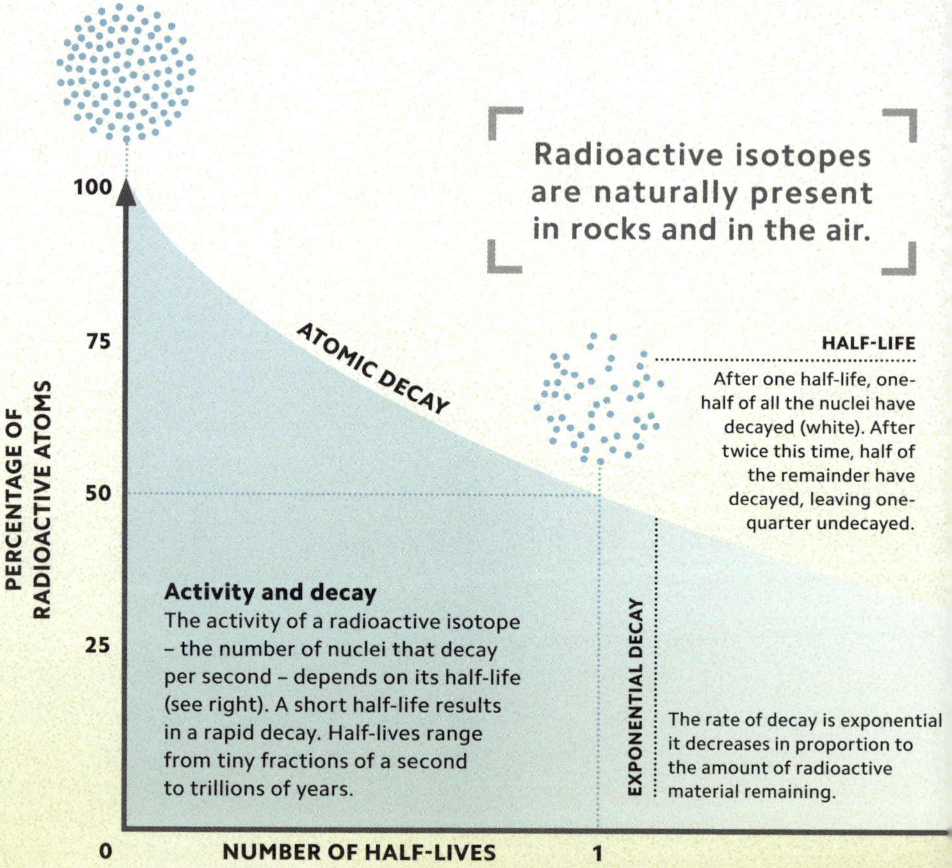

ATOMIC DECAY

PERCENTAGE OF RADIOACTIVE ATOMS

NUMBER OF HALF-LIVES

HALF-LIFE
After one half-life, one-half of all the nuclei have decayed (white). After twice this time, half of the remainder have decayed, leaving one-quarter undecayed.

Activity and decay
The activity of a radioactive isotope – the number of nuclei that decay per second – depends on its half-life (see right). A short half-life results in a rapid decay. Half-lives range from tiny fractions of a second to trillions of years.

EXPONENTIAL DECAY

The rate of decay is exponential it decreases in proportion to the amount of radioactive material remaining.

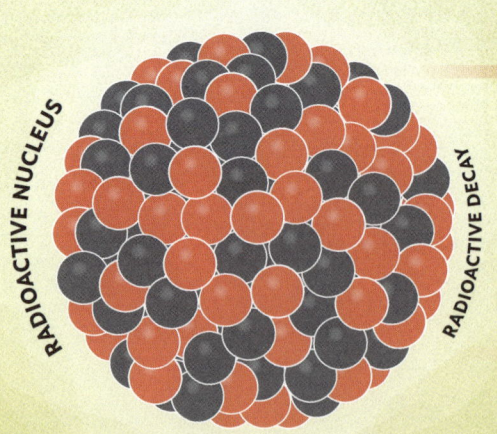

RADIOACTIVE NUCLEUS

RADIOACTIVE DECAY

Types of radiation
The different types of radiation emitted by decaying nuclei are characterized by their mass, energy, and how deeply they will penetrate. They can all damage atoms and molecules, leaving them ionized (see p.16); for this reason they are described as ionizing radiation.

NUCLEAR DECAY

The nuclei of some atoms are unstable because the forces holding their component particles together and those tending to push them apart are not in balance (see p.23). Such nuclei may spontaneously decay into more stable configurations, emitting energetic particles, such as alpha or beta particles, or gamma rays (see p.21) as they do so. We cannot say exactly when an individual nucleus will decay, only that there is a certain chance that it will decay in a particular time period. A sample of radioactive material decays rapidly at first, slowing as the number of undecayed nuclei decreases.

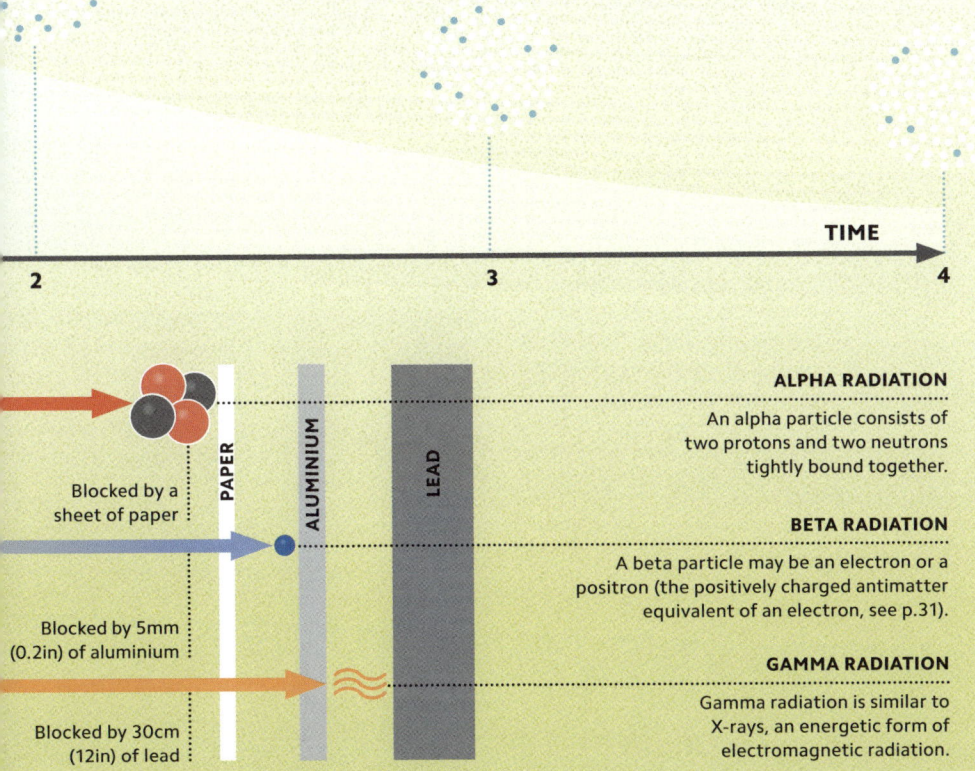

TIME

2   3   4

PAPER

ALUMINIUM

LEAD

Blocked by a
sheet of paper

Blocked by 5mm
(0.2in) of aluminium

Blocked by 30cm
(12in) of lead

ALPHA RADIATION

An alpha particle consists of two protons and two neutrons tightly bound together.

BETA RADIATION

A beta particle may be an electron or a positron (the positively charged antimatter equivalent of an electron, see p.31).

GAMMA RADIATION

Gamma radiation is similar to X-rays, an energetic form of electromagnetic radiation.

NEUTRON NUMBERS

Every atom of a specific element has the same number of protons in its nucleus (see p.14). However, the number of neutrons can vary so each element exists in more than one form – or isotope – each with a different mass. Some isotopes of an element are stable, while others are unstable and thus radioactive (see pp.20–21); these are called radioisotopes. Some elements, such as uranium, exist only in unstable form.

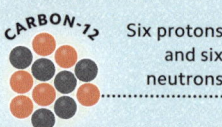

Six protons and six neutrons

STABLE

Six protons and seven neutrons

STABLE

Six protons and eight neutrons

UNSTABLE, RADIOACTIVE

Carbon isotopes

Carbon occurs naturally in three isotopes; 99 per cent is carbon-12 (C-12), about 1 per cent is C-13, and one out of every trillion carbon atoms is C-14.

Carbon dating

The carbon atoms in a living organism are replenished as it respires, so the proportions of C-12 to C-13 to C-14 in its body remain constant. After death, carbon is no longer replenished. As C-14 undergoes radioactive decay, its proportion in the animal's remains decreases. Measuring the proportions of carbon isotopes in the remains allows scientists to calculate their age.

> Isotopes of an element have near identical chemical properties.

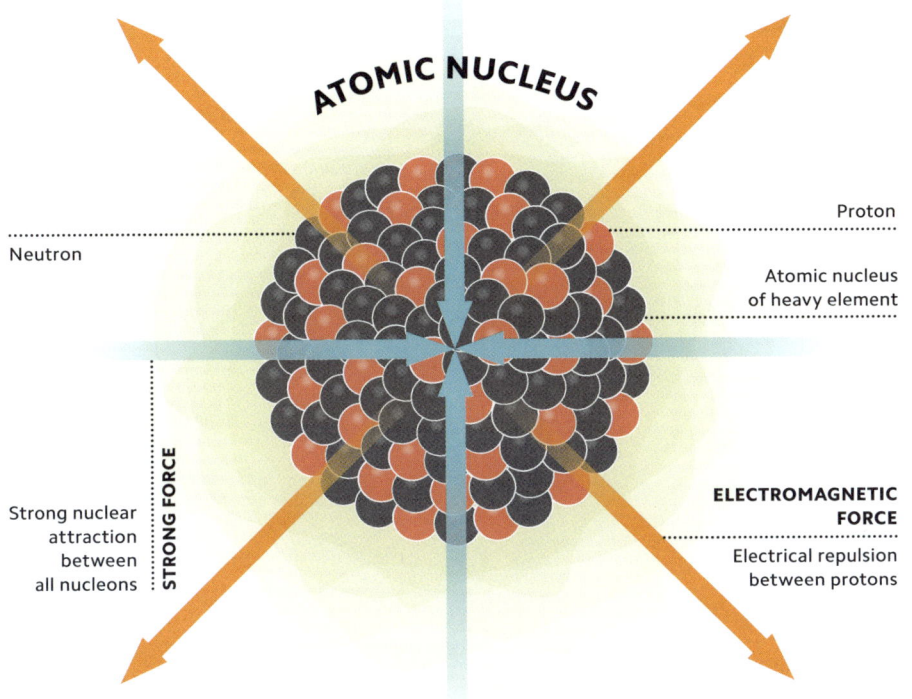

ATOMIC NUCLEUS

Proton

Neutron

Atomic nucleus
of heavy element

STRONG FORCE

ELECTROMAGNETIC
FORCE

Strong nuclear
attraction
between
all nucleons

Electrical repulsion
between protons

NUCLEAR BALANCE

Atomic nuclei contain protons (a hydrogen atom has just one). Since protons have positive electrical charge they all repel one another. Known as the electromagnetic force, this tends to break the nucleus apart. A second force – the strong force – acts to hold the nucleus together. It is an attractive force that acts between all nucleons (protons and neutrons) to counteract the electrical repulsion. The electromagnetic force and the strong force are two of the fundamental forces of nature (see pp.32–33). For a nucleus to be stable, these forces must be balanced.

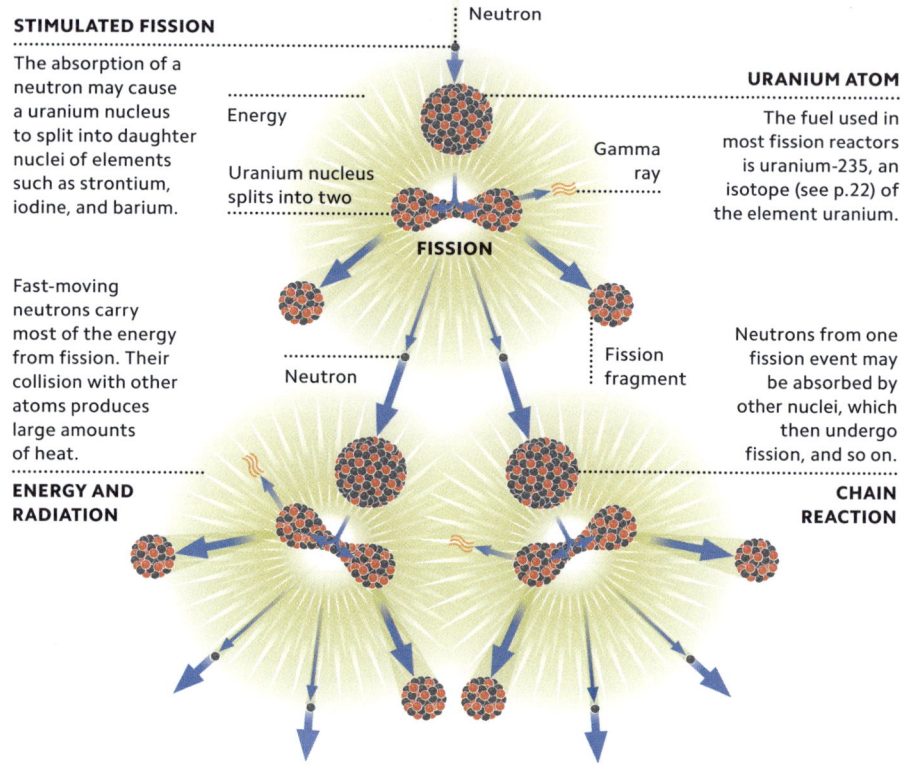

STIMULATED FISSION

The absorption of a neutron may cause a uranium nucleus to split into daughter nuclei of elements such as strontium, iodine, and barium.

Fast-moving neutrons carry most of the energy from fission. Their collision with other atoms produces large amounts of heat.

ENERGY AND RADIATION

Neutron

Energy

Uranium nucleus splits into two

FISSION

Neutron

URANIUM ATOM

The fuel used in most fission reactors is uranium-235, an isotope (see p.22) of the element uranium.

Gamma ray

Fission fragment

Neutrons from one fission event may be absorbed by other nuclei, which then undergo fission, and so on.

CHAIN REACTION

Large atomic nuclei may undergo nuclear fission, in which they split into two smaller nuclei, releasing a small number of neutrons in the process. This may happen spontaneously in unstable nuclei of high mass. Fission may also be triggered deliberately to produce energy in nuclear reactors, or in the destructive explosion of nuclear weapons. In a reactor, elements such as uranium or plutonium are hit with neutrons travelling at the correct speed to trigger a chain reaction. The reaction releases high-speed particles and produces energetic gamma rays.

S P L I T T I N G A P A R T

POWER FROM THE NUCLEUS

Nuclear power stations harness energy produced by nuclear fission.
Their design allows the nuclear chain reaction (see left) to be precisely
controlled: a moderator chemical surrounds the fuel, slowing neutrons
to the speed required to trigger fission, while control rods can be
deployed to absorb excess neutrons and so prevent a runaway
reaction. There are more than 400 commercial reactors
in operation around the world.

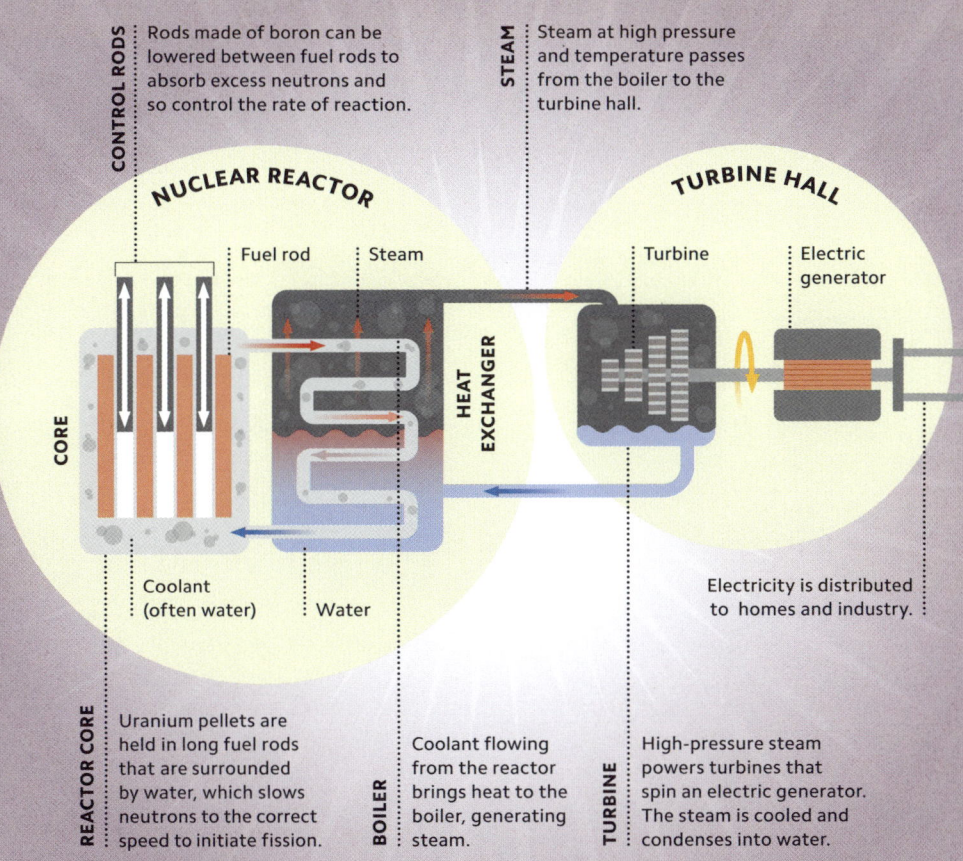

CONTROL RODS
Rods made of boron can be lowered between fuel rods to absorb excess neutrons and so control the rate of reaction.

STEAM
Steam at high pressure and temperature passes from the boiler to the turbine hall.

NUCLEAR REACTOR

Fuel rod Steam

HEAT EXCHANGER

CORE

Coolant (often water) Water

TURBINE HALL

Turbine Electric generator

Electricity is distributed to homes and industry.

REACTOR CORE
Uranium pellets are held in long fuel rods that are surrounded by water, which slows neutrons to the correct speed to initiate fission.

BOILER
Coolant flowing from the reactor brings heat to the boiler, generating steam.

TURBINE
High-pressure steam powers turbines that spin an electric generator. The steam is cooled and condenses into water.

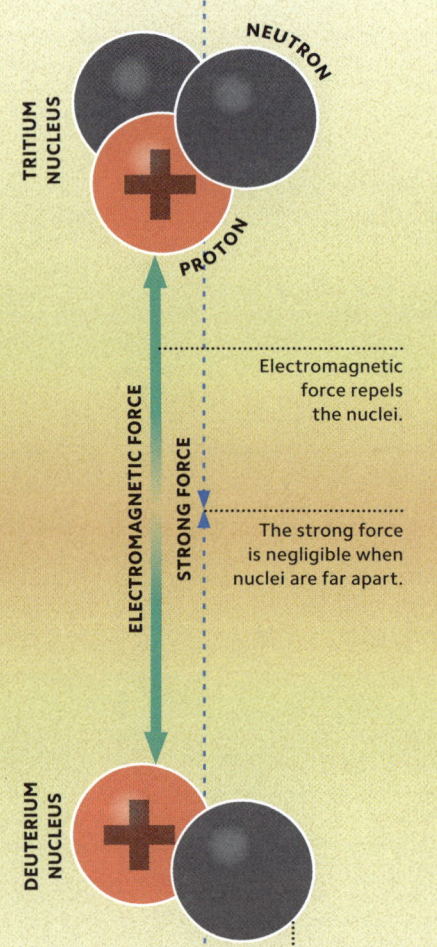

TRITIUM
NUCLEUS

NEUTRON

PROTON

ELECTROMAGNETIC FORCE

STRONG FORCE

Electromagnetic
force repels
the nuclei.

The strong force
is negligible when
nuclei are far apart.

DEUTERIUM
NUCLEUS

LOW-SPEED
NUCLEI

At low temperatures, atomic
nuclei move slowly. Electrical
repulsion between nuclei of two
hydrogen isotopes – deuterium
(H-2) and tritium (H-3), which
both carry a positive charge
– means that they are held apart.

WHEN NUCLEI MERGE

When two small nuclei come close together, they may merge. This is nuclear fusion, a process that takes place in stars. Conditions must be right. If the nuclei are moving too slowly, the electrical repulsion between them will be too great for fusion to occur. If they are moving too fast, they simply rush past one another. However, at an intermediate speed, the two nuclei may come close enough together for the strong nuclear force to overcome the electrical repulsion between them and they fuse. As with nuclear fission (see p.24), energy is released and carried off by fast-moving neutrons.

STRONG FORCE

HIGH-SPEED NUCLEI

At high temperatures, atomic nuclei move rapidly and come into close proximity. The strong force brings the nuclei together, overcoming electrical repulsion; fusion occurs when nuclei collide at the right speed.

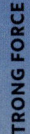

ELECTROMAGNETIC FORCE

STRONG FORCE

The two hydrogen nuclei merge briefly.

NEUTRON

FAST NEUTRON

A neutron moving at one-sixth of the speed of light is released. It carries off most of the energy released in the fusion process.

HELIUM NUCLEUS

From hydrogen to helium

If the conditions are right, the nuclei of deuterium and tritium can fuse to produce a helium-4 nucleus. Fusion requires the nuclei to be moving fast. Temperatures of over 100 million °C are needed to achieve these speeds. A fast-moving neutron is released in the process.

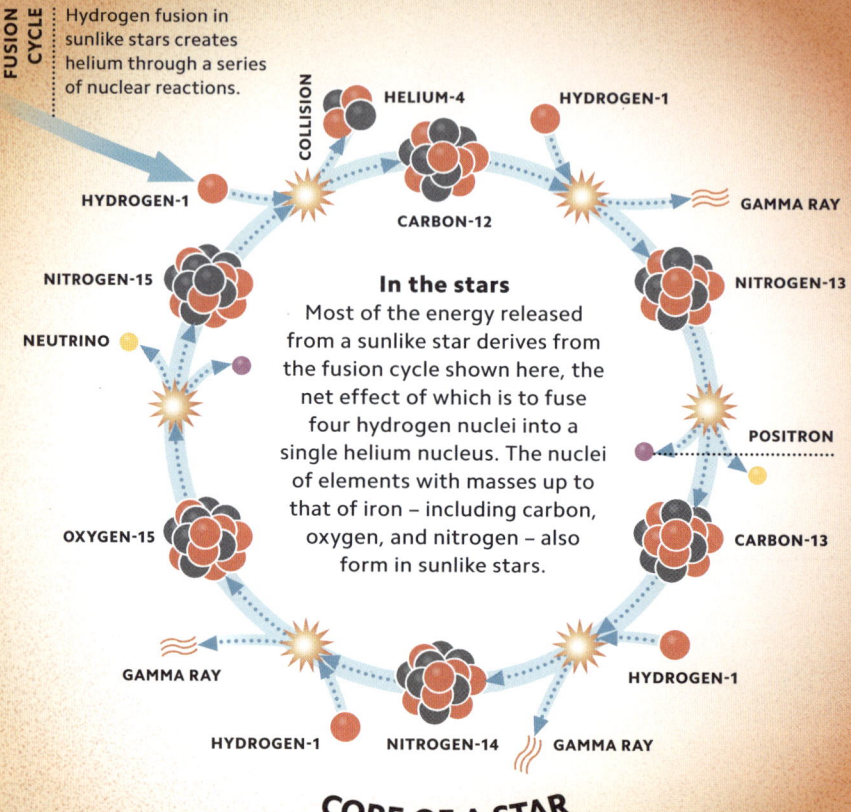

FUSION CYCLE

Hydrogen fusion in sunlike stars creates helium through a series of nuclear reactions.

HELIUM-4

HYDROGEN-1

COLLISION

CARBON-12

GAMMA RAY

HYDROGEN-1

NITROGEN-13

NITROGEN-15

NEUTRINO

In the stars

Most of the energy released from a sunlike star derives from the fusion cycle shown here, the net effect of which is to fuse four hydrogen nuclei into a single helium nucleus. The nuclei of elements with masses up to that of iron – including carbon, oxygen, and nitrogen – also form in sunlike stars.

POSITRON

OXYGEN-15

CARBON-13

GAMMA RAY

HYDROGEN-1

HYDROGEN-1

NITROGEN-14

GAMMA RAY

CORE OF A STAR

ELEMENT FACTORIES

The formation of atomic nuclei is called nucleosynthesis. Nuclei of the lightest elements – hydrogen, helium, and lithium – were created minutes after the Big Bang (see p.151) when tiny particles combined into protons and neutrons. Formation of heavier nuclei began some 500 million years later, as a result of nuclear fusion within large stars, and during their deaths, when huge stars exploded as supernovas (see p.147).

Much of our knowledge of the nature of matter comes from experiments in which charged particles, such as protons, are smashed together at near-light speed. These high-energy collisions produce exotic particles, such as quarks, leptons, and bosons (see pp.34–35), providing vital clues to the underlying nature of matter.

SMASHING STUFF

BEAM INJECTION

Two beams of high-energy particles, travelling in opposite directions, are injected into the circular tube.

Powerful electromagnets focus the particle beams and force them to follow a circular path.

FOCUSING THE BEAMS

Particle collider

A particle accelerator is a tube many kilometres in diameter. It contains a vacuum. Particles are accelerated to high energies around this circular track and forced to collide. Detectors track the resulting particles.

VACUUM CHAMBER

Detectors are placed where particles in the two beams are forced to collide.

DETECTING COLLISIONS

ACCELERATION

As the particles circulate, they are accelerated to close to the speed of light by strong electric fields.

COSMIC
RAYS

30,000M/100,000FT

INITIAL COLLISION

A cosmic ray hits an atomic nucleus high in the atmosphere. Protons, electrons, and pions are produced in the high-energy collision.

MUON FORMATION

Charged pions decay within a fraction of a second, forming leptons, such as muons (see p.35).

RADIATION

Neutral pions decay and produce gamma radiation, which interacts with more nuclei, forming electron–positron pairs.

HADRON SHOWERS

Some pions collide with other nuclei, producing a shower of hadrons (see pp.34–35).

DETECTOR

The cascade of particles can be detected by instruments on the ground or aboard aircraft.

PARTICLE ZOO

Our planet is bombarded by trillions of cosmic rays every day. Although called rays, they are in fact particles – bare atomic nuclei – originating from the Sun and far beyond the Solar System, and travelling at close to the speed of light. Most are deflected by Earth's magnetic field, but some collide with molecules in the atmosphere, producing showers of particles that rain down on Earth. Many of these particles are unstable and decay rapidly; by analysing these decay products, scientists have discovered much about the nature of the original particles. Antimatter (see opposite) and muons are part of the exotic "particle zoo" exposed by cosmic rays.

MATTER IN A MIRROR

The world that we can see and touch is made of matter. However, for every type of known particle – such as the electron and proton – there is a corresponding antiparticle with opposite properties, such as charge and spin. Certain types of antimatter are common; for example, positrons (antielectrons) are produced in the beta decay of some radioactive elements. However, there seems to be much less antimatter than matter in the Universe – an observation that remains unexplained.

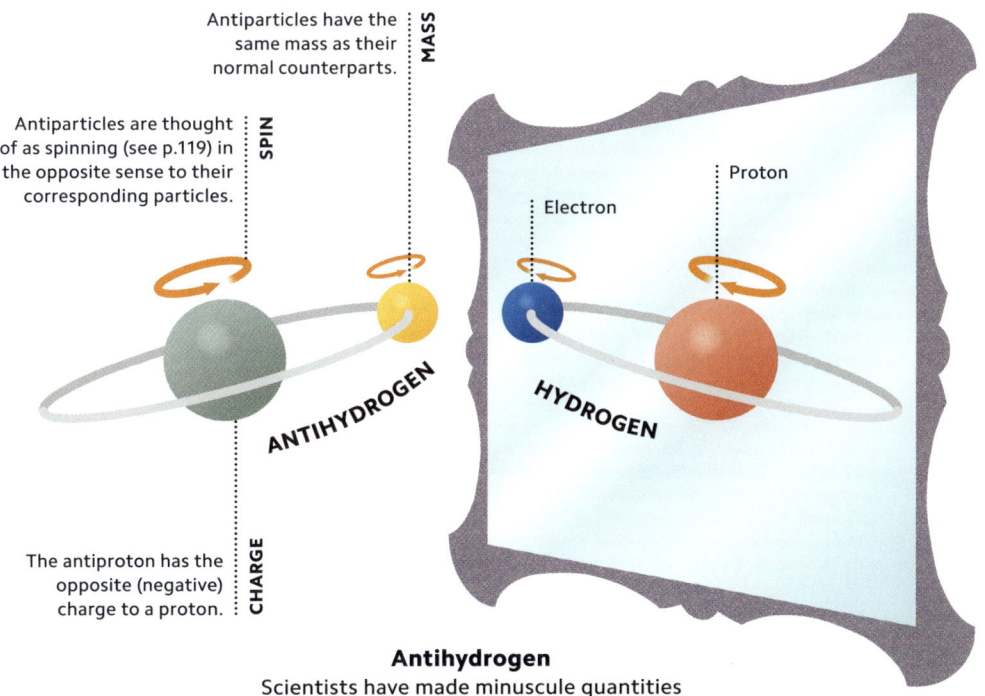

Antiparticles have the same mass as their normal counterparts.

MASS

SPIN

Antiparticles are thought of as spinning (see p.119) in the opposite sense to their corresponding particles.

Proton

Electron

ANTIHYDROGEN

HYDROGEN

CHARGE

The antiproton has the opposite (negative) charge to a proton.

Antihydrogen
Scientists have made minuscule quantities of antihydrogen. Matter and antimatter particles are always produced as a pair. When they meet, they annihilate each other in a burst of energy.

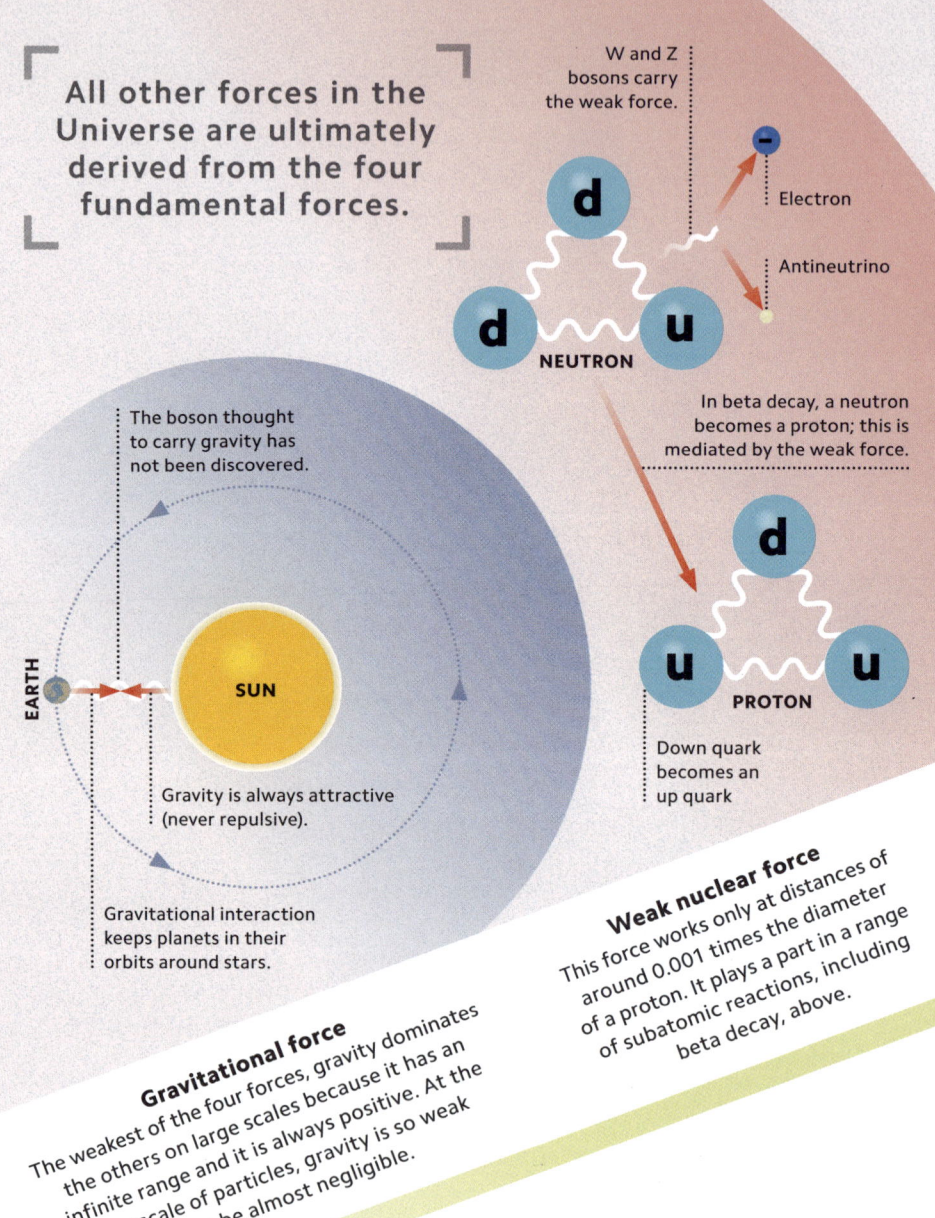

All other forces in the Universe are ultimately derived from the four fundamental forces.

W and Z bosons carry the weak force.

−

Electron

Antineutrino

d

d   **u**

NEUTRON

In beta decay, a neutron becomes a proton; this is mediated by the weak force.

The boson thought to carry gravity has not been discovered.

d

u   **u**

PROTON

Down quark becomes an up quark

EARTH

SUN

Gravity is always attractive (never repulsive).

Gravitational interaction keeps planets in their orbits around stars.

Weak nuclear force
This force works only at distances of around 0.001 times the diameter of a proton. It plays a part in a range of subatomic reactions, including beta decay, above.

Gravitational force
The weakest of the four forces, gravity dominates the others on large scales because it has an infinite range and it is always positive. At the tiny scale of particles, gravity is so weak as to be almost negligible.

WEAK

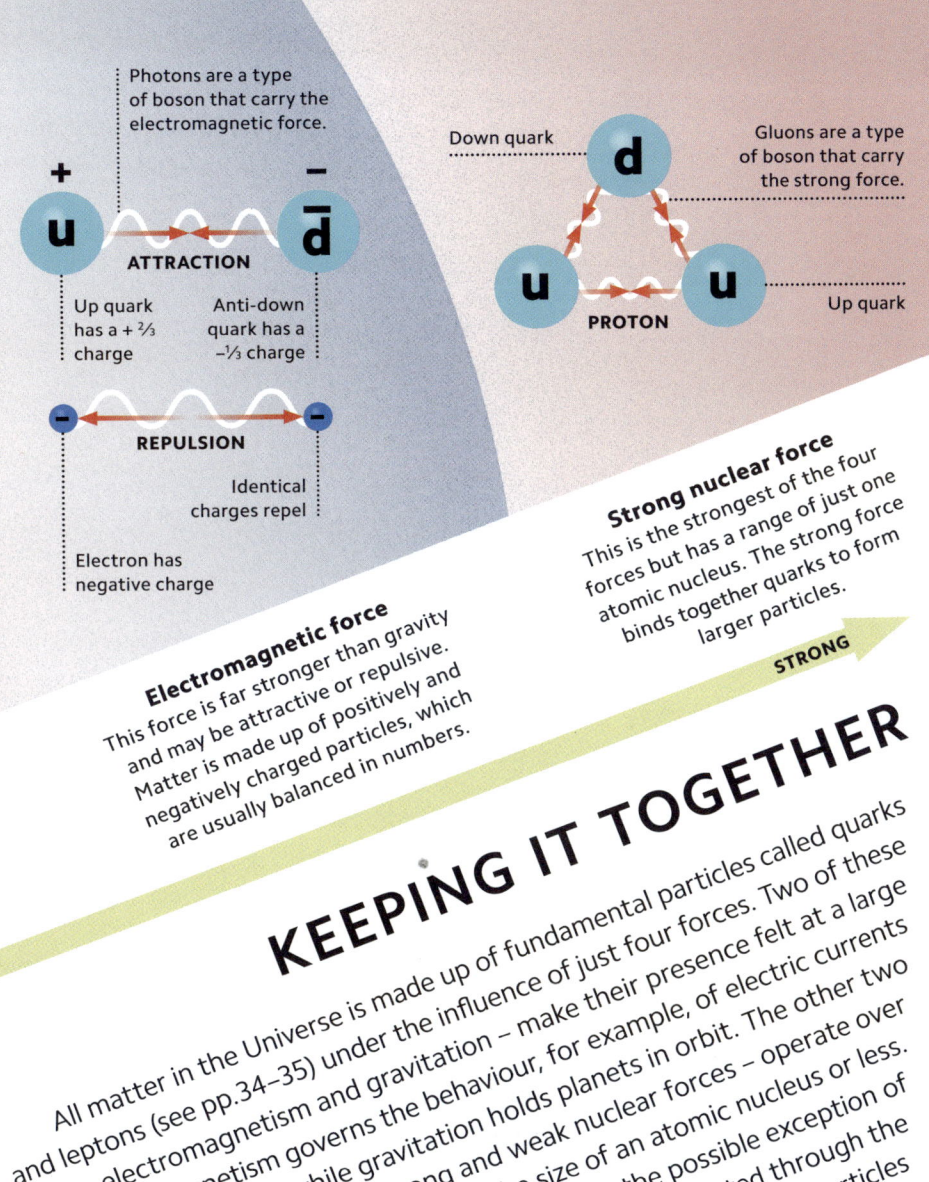

Photons are a type of boson that carry the electromagnetic force.

+ u

‒ d̄

ATTRACTION

Up quark has a + ⅔ charge

Anti-down quark has a −⅓ charge

‒ ←→ ‒

REPULSION

Identical charges repel

Electron has negative charge

Down quark

Gluons are a type of boson that carry the strong force.

d

u u

PROTON

Up quark

Electromagnetic force
This force is far stronger than gravity and may be attractive or repulsive. Matter is made up of positively and negatively charged particles, which are usually balanced in numbers.

Strong nuclear force
This is the strongest of the four forces but has a range of just one atomic nucleus. The strong force binds together quarks to form larger particles.

STRONG

KEEPING IT TOGETHER

All matter in the Universe is made up of fundamental particles called quarks and leptons (see pp.34–35) under the influence of just four forces. Two of these forces – electromagnetism and gravitation – make their presence felt at a large scale. Electromagnetism governs the behaviour, for example, of electric currents powering our homes, while gravitation holds planets in orbit. The other two forces – called the strong and weak nuclear forces – operate over minute distances, the size of an atomic nucleus or less. These forces (with the possible exception of gravity) are exerted through the exchange of carrier particles called bosons.

Quark combinations

Quarks never exist on their own but always in combination with other quarks in particles called hadrons. These may be composed of two, three, or four quarks or antiquarks, (see p.31). The properties of a hadron – such as its charge – derive from the properties of its constituent quarks.

BARYONS

Baryons consist of three quarks. Protons and neutrons are examples of baryons; a proton is a stable baryon made from two up quarks and one down quark; a neutron has one up and two down quarks.

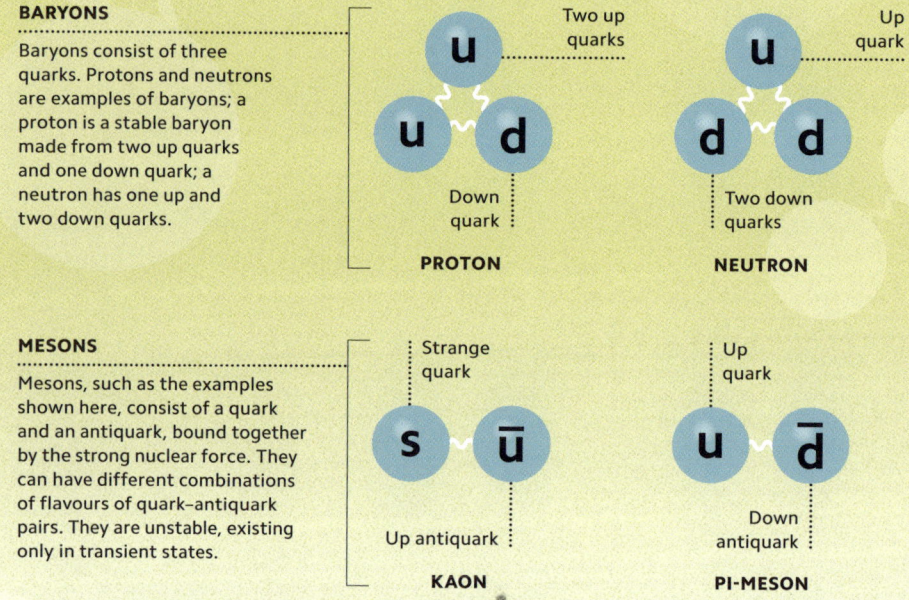

Two up quarks

Down quark

PROTON

Up quark

Two down quarks

NEUTRON

MESONS

Mesons, such as the examples shown here, consist of a quark and an antiquark, bound together by the strong nuclear force. They can have different combinations of flavours of quark–antiquark pairs. They are unstable, existing only in transient states.

Strange quark

Up antiquark

KAON

Up quark

Down antiquark

PI-MESON

GETTING FUNDAMENTAL

The protons and neutrons that make up atomic nuclei are themselves composed of smaller particles called quarks. As far as is known, these are elementary particles that cannot be further broken down. Electrons and neutrinos belong to another class of elementary particles called leptons. There are six types of quark – up, down, charm, strange, top, and bottom. Their names have arisen historically and tell us little about their nature. Quarks are affected by all four fundamental forces (see pp.32–33).

ALL THERE IS

Our knowledge know about elementary particles and the fundamental forces that act upon them is summed up in what is called the standard model. This sets out the six types of quark and the six types of lepton. A corresponding antiparticle (see p.29) exists for each of these particles. In addition, the standard model includes the five gauge bosons that transfer the fundamental forces that act on matter (see pp.32–33). Gravity is not included in the standard model.

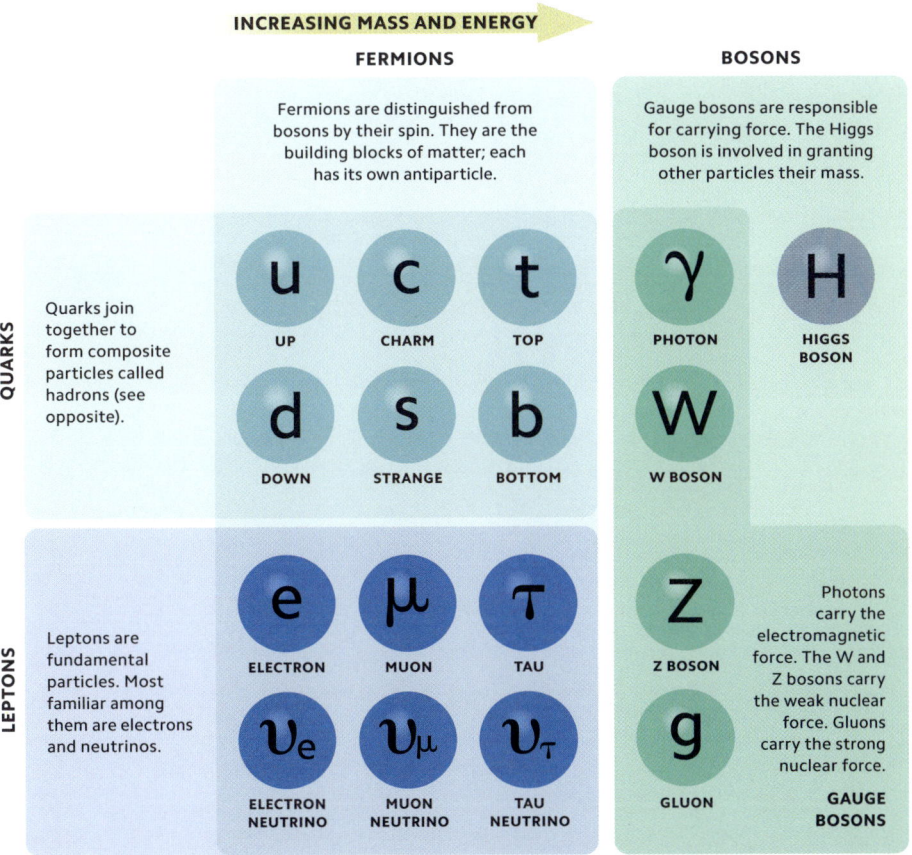

INCREASING MASS AND ENERGY

FERMIONS

BOSONS

Fermions are distinguished from bosons by their spin. They are the building blocks of matter; each has its own antiparticle.

Gauge bosons are responsible for carrying force. The Higgs boson is involved in granting other particles their mass.

QUARKS

Quarks join together to form composite particles called hadrons (see opposite).

u	c	t
UP	CHARM	TOP
d	s	b
DOWN	STRANGE	BOTTOM

γ — PHOTON   H — HIGGS BOSON

W — W BOSON

LEPTONS

Leptons are fundamental particles. Most familiar among them are electrons and neutrinos.

e	μ	τ
ELECTRON	MUON	TAU
ν_e	ν_μ	ν_τ
ELECTRON NEUTRINO	MUON NEUTRINO	TAU NEUTRINO

Z — Z BOSON

g — GLUON

Photons carry the electromagnetic force. The W and Z bosons carry the weak nuclear force. Gluons carry the strong nuclear force.

GAUGE BOSONS

FORCE
MOTIO

S A N D
N

Applying a force to an object may cause it to move or change shape or size. Explaining – and being able to predict – the outcomes of these encounters is one of the most fundamental aspects of physics. It has allowed us to understand the motion of the planets and stars, and to design the machines and materials that have transformed the world. Isaac Newton's three laws of motion form the basis of classical physics. However, these laws make the assumption that the concepts of distance, time, and mass are absolute. These assumptions fail when objects move very fast or are extremely small. These areas are addressed in the modern fields of relativity and quantum mechanics.

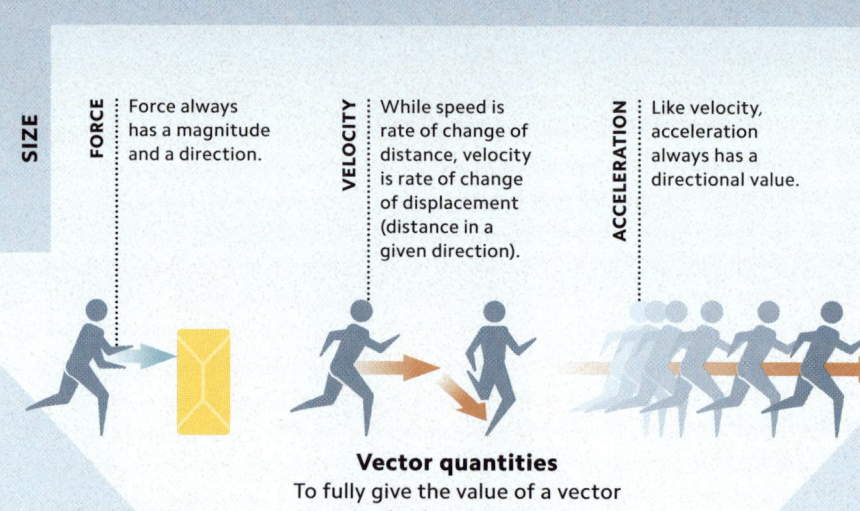

SIZE ONLY

MASS
Mass is a property of all matter. It is a measure of an object's resistance to movement when a force is applied.

TEMPERATURE
Temperature is a measure of the average movement (kinetic) energy of all the particles in a body.

DISTANCE
This is a measure of how far a body moves along any path.

KG

A   B

Scalar quantities

A quantity whose value is measured against a scale of units (such as kilograms, degrees Celsius, or metres) without any direction is known as a scalar quantity.

SIZE

FORCE
Force always has a magnitude and a direction.

VELOCITY
While speed is rate of change of distance, velocity is rate of change of displacement (distance in a given direction).

ACCELERATION
Like velocity, acceleration always has a directional value.

Vector quantities

To fully give the value of a vector quantity, both its magnitude and direction must be stated.

DIRECTION

Combining vectors

Vectors can be added to calculate the outcome when forces, or velocities, are combined. This has real-world uses, such as working out the force in newtons (N) of two tug boats towing a large ship.

SIZE AND DIRECTION

Physics involves the measurement of many different quantities. Some, such as temperature, distance, speed, mass, and energy, have only magnitude (size). These are called scalar quantities. Others, called vector quantities, have direction as well as magnitude. They include displacement, velocity, and acceleration (distance, speed, and rate of change of speed in a particular direction), as well as force. Vectors can be added to find a combined vector. Conversely, the effects of a single vector can be broken down into two or more components.

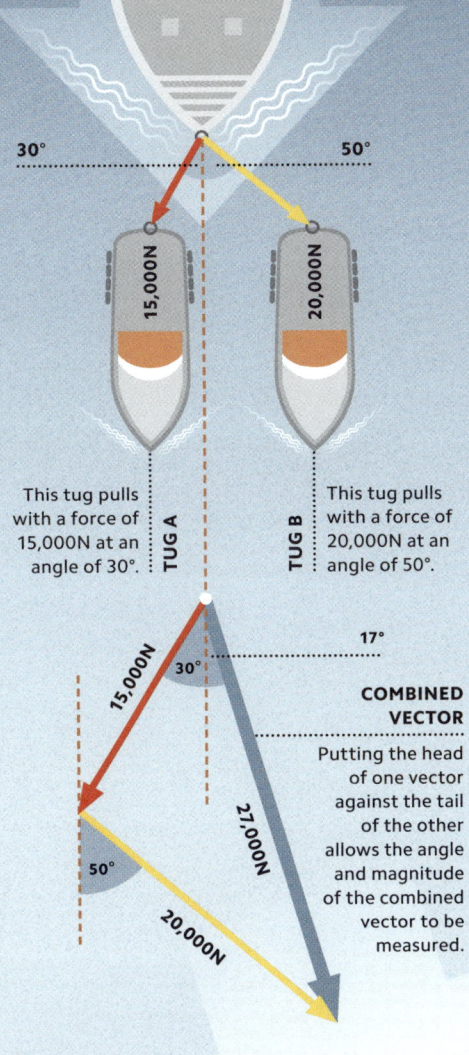

SHIP

30°   50°

15,000N   20,000N

This tug pulls with a force of 15,000N at an angle of 30°.

TUG A

TUG B

This tug pulls with a force of 20,000N at an angle of 50°.

15,000N   30°   17°

COMBINED VECTOR

Putting the head of one vector against the tail of the other allows the angle and magnitude of the combined vector to be measured.

50°

27,000N

20,000N

PRACTICAL MECHANICS

Isaac Newton's three laws of motion describe and predict the relationship between the motion of an object and the forces acting upon it. Newton's laws are fundamental to classical mechanics, explaining, for example, the motion of the planets. However, they do not apply at speeds approaching the speed of light or to the behaviour of subatomic particles, for which the theories of relativity (see pp.126–137) and quantum physics (see pp.108–25) are required.

Newton's first law
An object remains at rest or in uniform motion unless acted upon by a force. So a stationary ice hockey puck will remain stationary, and one gliding on ice at a constant speed will continue to slide forever, unless it meets an external force.

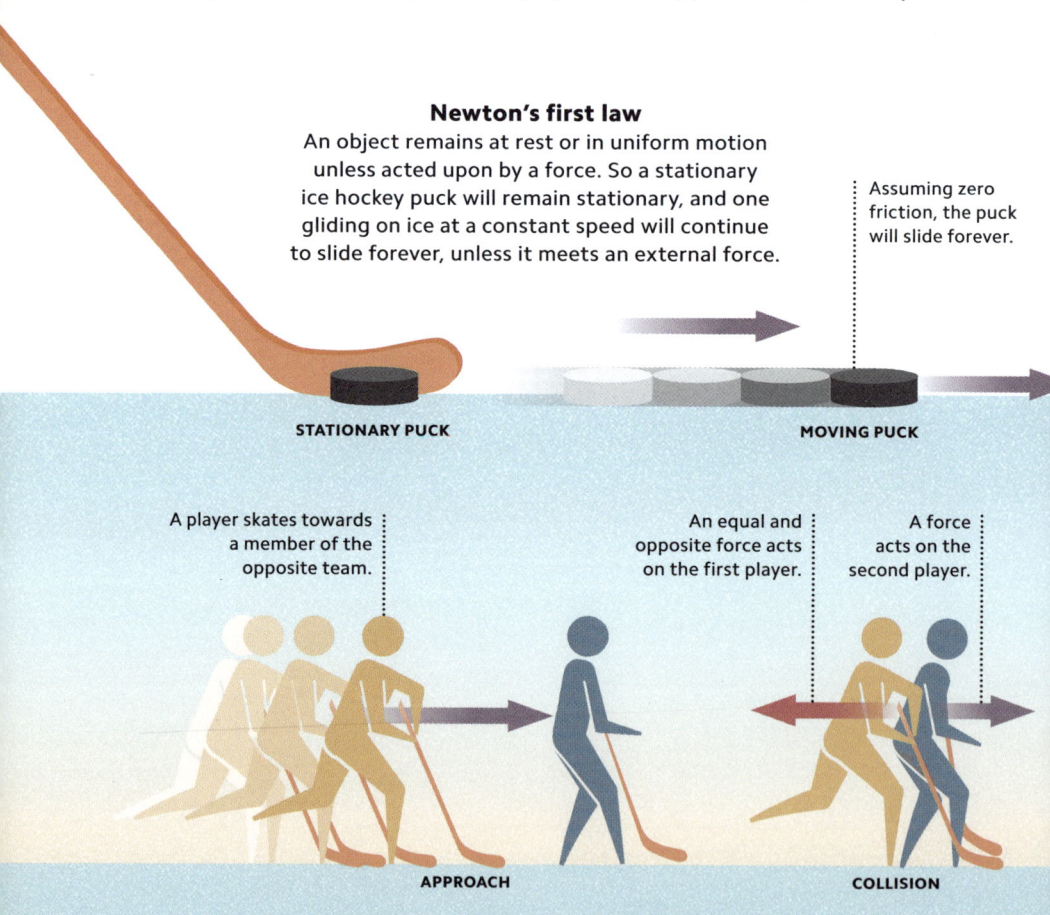

Assuming zero friction, the puck will slide forever.

STATIONARY PUCK

MOVING PUCK

A player skates towards a member of the opposite team.

An equal and opposite force acts on the first player.

A force acts on the second player.

APPROACH

COLLISION

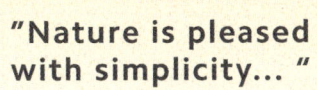

"Nature is pleased
with simplicity... "
Isaac Newton

Newton's second law

When a constant force acts on a
mass, it causes that mass to accelerate.
The acceleration increases as the force
increases or as the mass decreases.
Acceleration is a change of velocity
or a change of direction, because
velocity is a vector quantity
(see pp.38–39).

As the hockey stick
strikes the puck,
it causes it to
accelerate.

If the stick deflects the
puck, it causes it to change
direction, and so accelerate.

Mass   Acceleration

Net force

CHANGING VELOCITY

**CHANGING
VELOCITY**

The first player
decelerates.

The second player
accelerates.

Newton's third law

For every action, there
is an equal and opposite
reaction. So if an ice hockey
player pushes another, both
will accelerate. Any force
applied is always matched by
an equal and opposite force.

ACCELERATION

Elastic and inelastic collisions

In an elastic collision the total kinetic energy of the objects after the collision is unchanged. In an inelastic collision, some becomes heat or sound. In each case, the total momentum of the system is conserved. Collisions between snooker balls are close to being elastic.

GREEN BALL

The ball has some momentum in the forward direction and some in the leftward direction.

INITIAL FORCE

Force applied by the cue accelerates the cue ball.

The green ball is initially at rest.

The balls collide obliquely.

FINAL MOMENTUM

The lateral momentum of the two balls is exactly equal and opposite; net momentum is in the forward direction.

INITIAL MOMENTUM

The cue ball's momentum is all in the forward direction and is given by its mass multiplied by its velocity.

CUE BALL

The cue ball has some momentum in the forward direction and some in the rightward direction.

QUANTITIES OF MOTION

Momentum is the amount of motion that an object possesses; numerically, it is the mass of the object multiplied by its velocity. As such, it is a vector quantity (see p.38). Newton's third law (see p.41) tells us that within a closed system, such as a collision between objects, momentum is always conserved. This means that the total momentum of the colliding objects does not change; it is just redistributed between the objects.

Collision of unequal masses

When two objects of different masses collide, the object with higher mass changes velocity less than the object with lower mass. This explains why it is better to be in a heavier car in the event of a collision.

A rotating body is said to have rotational momentum, another conserved quantity.

INITIAL MOMENTUM
This ball's momentum is all in the forward direction and is given by its mass multiplied by its velocity.

STATIONARY BALL
The stationary yellow ball has twice the mass of the blue ball.

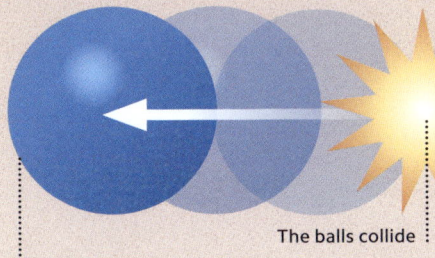

The balls collide

BACKWARD MOMENTUM
The smaller ball bounces back in a collision with the larger ball. It has momentum in the backward direction.

MOMENTUM CONSERVED
The overall momentum of the system is the same as that before the collision.

FORWARD MOMENTUM
The more massive ball has forward momentum after the collision.

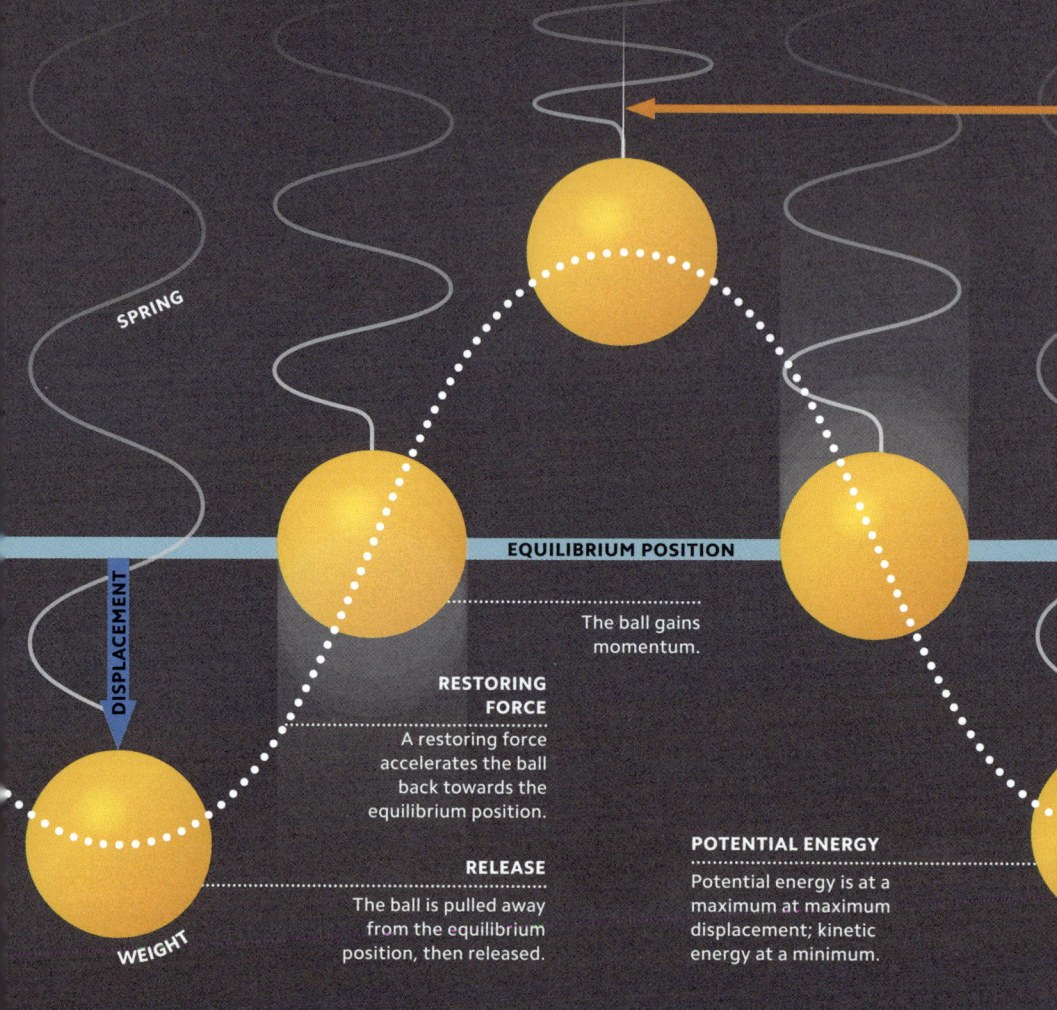

SPRING

EQUILIBRIUM POSITION

DISPLACEMENT

The ball gains momentum.

RESTORING FORCE
A restoring force accelerates the ball back towards the equilibrium position.

RELEASE
The ball is pulled away from the equilibrium position, then released.

POTENTIAL ENERGY
Potential energy is at a maximum at maximum displacement; kinetic energy at a minimum.

WEIGHT

Mass on a spring

A ball on a spring exhibits SHM when it is pulled away from its equilibrium position and allowed to oscillate. Its frequency remains the same, regardless of the initial displacement of the ball. It depends only on the mass of the ball and the stiffness of the spring.

SHM describes both the swinging of a clock pendulum and the microscopic vibration of molecules.

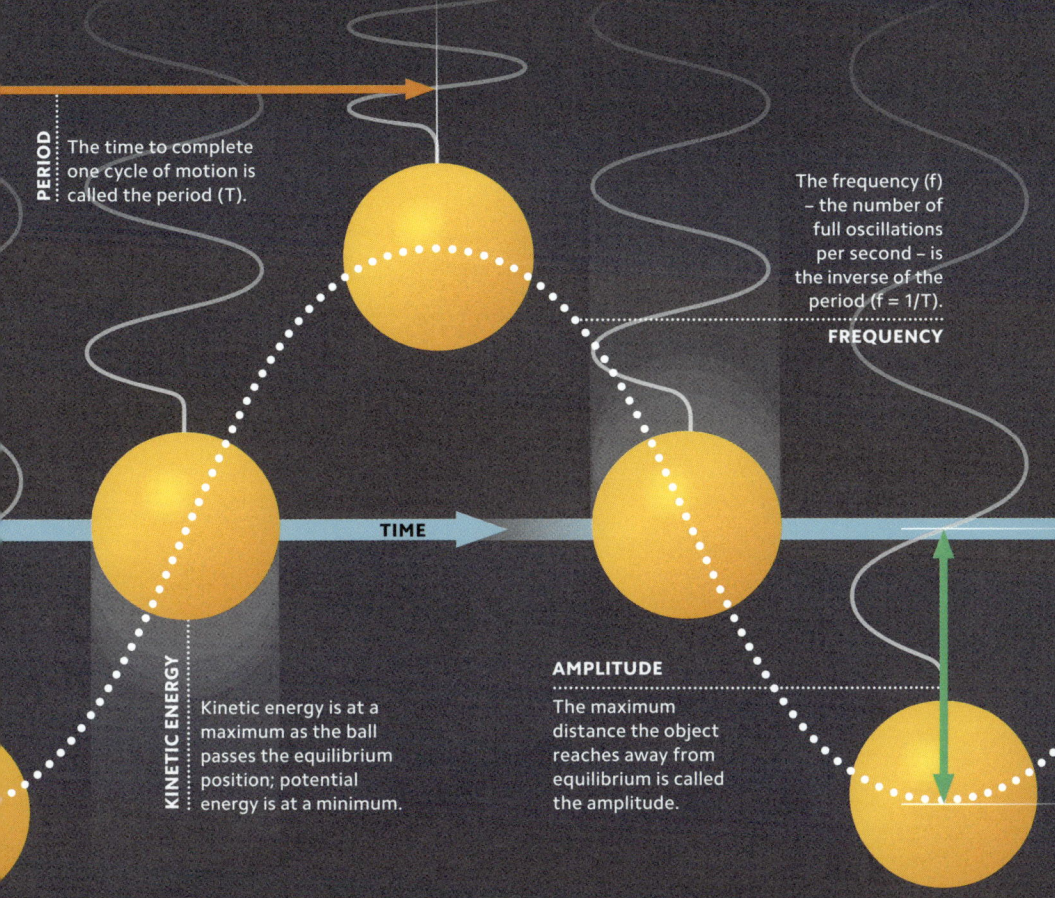

The time to complete one cycle of motion is called the period (T).

The frequency (f) – the number of full oscillations per second – is the inverse of the period (f = 1/T).
FREQUENCY

TIME

KINETIC ENERGY
Kinetic energy is at a maximum as the ball passes the equilibrium position; potential energy is at a minimum.

AMPLITUDE
The maximum distance the object reaches away from equilibrium is called the amplitude.

MOVING IN RHYTHM

Simple harmonic motion (SHM) is a periodic (regular, repeating) to and fro motion. It is the phenomenon behind the repeating vibrations that create waves such as sound, water, and radio waves. In SHM, there is an equilibrium position at which the oscillating object experiences no net force and remains at rest if not disturbed. If it is moved away from that point, the object is pulled back by a restoring force that is proportional to the distance from equilibrium. This force repeatedly pulls the object back towards equilibrium, but its momentum carries it onwards. The force reverses to pull the object back again.

ROUND AND ROUND

Newton's second law (see pp.40–41) tells us that a force is needed to change the direction of a moving object. An object that is moving round in a circle is constantly changing direction, and so needs a constant force. That force, called a centripetal force, acts towards the centre of the circle. For an object whirling on a string, the centripetal force is supplied by tension in the string. For a planet in orbit around the Sun, or a satellite around Earth, it is gravitational force.

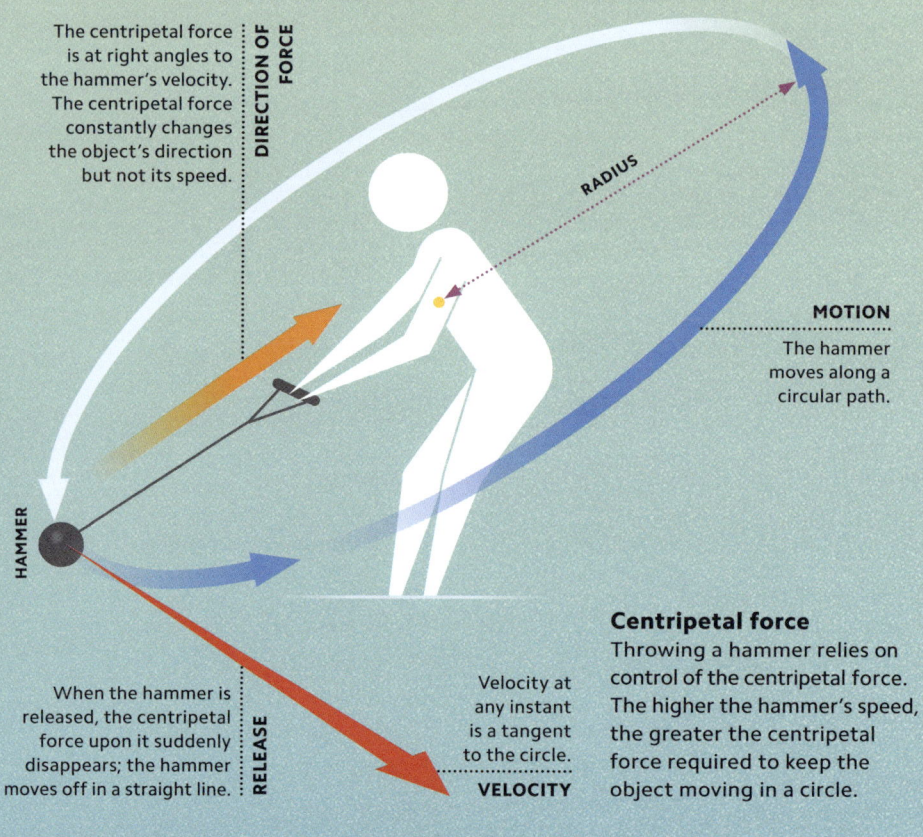

The centripetal force is at right angles to the hammer's velocity. The centripetal force constantly changes the object's direction but not its speed.

DIRECTION OF FORCE

RADIUS

MOTION
The hammer moves along a circular path.

HAMMER

When the hammer is released, the centripetal force upon it suddenly disappears; the hammer moves off in a straight line.

RELEASE

Velocity at any instant is a tangent to the circle.

VELOCITY

Centripetal force
Throwing a hammer relies on control of the centripetal force. The higher the hammer's speed, the greater the centripetal force required to keep the object moving in a circle.

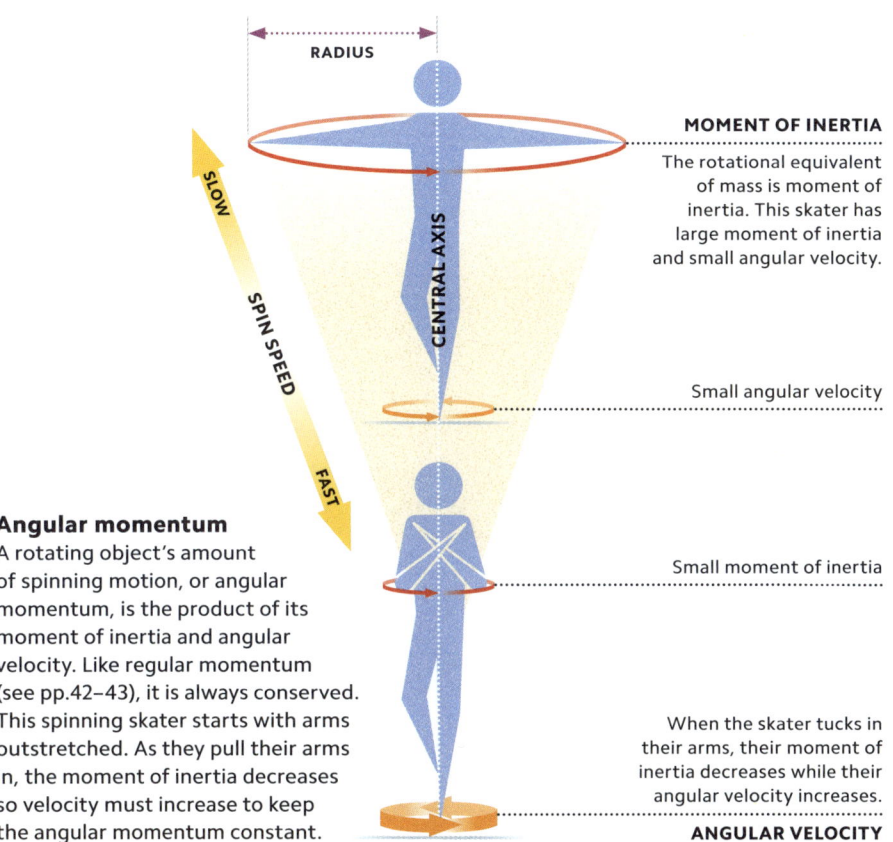

RADIUS

MOMENT OF INERTIA
The rotational equivalent of mass is moment of inertia. This skater has large moment of inertia and small angular velocity.

SLOW

SPIN SPEED

CENTRAL AXIS

FAST

Small angular velocity

Angular momentum
A rotating object's amount of spinning motion, or angular momentum, is the product of its moment of inertia and angular velocity. Like regular momentum (see pp.42–43), it is always conserved. This spinning skater starts with arms outstretched. As they pull their arms in, the moment of inertia decreases so velocity must increase to keep the angular momentum constant.

Small moment of inertia

When the skater tucks in their arms, their moment of inertia decreases while their angular velocity increases.

ANGULAR VELOCITY

IN A WHIRL

Rotational motion, when an object spins in a circle around an axis, occurs when a force is applied at a distance from the axis, creating a turning force, or torque. The magnitude of the torque is proportional to both the magnitude of the force and its distance from the axis. Just as force in linear motion changes velocity, torque changes angular velocity in a rotating system.

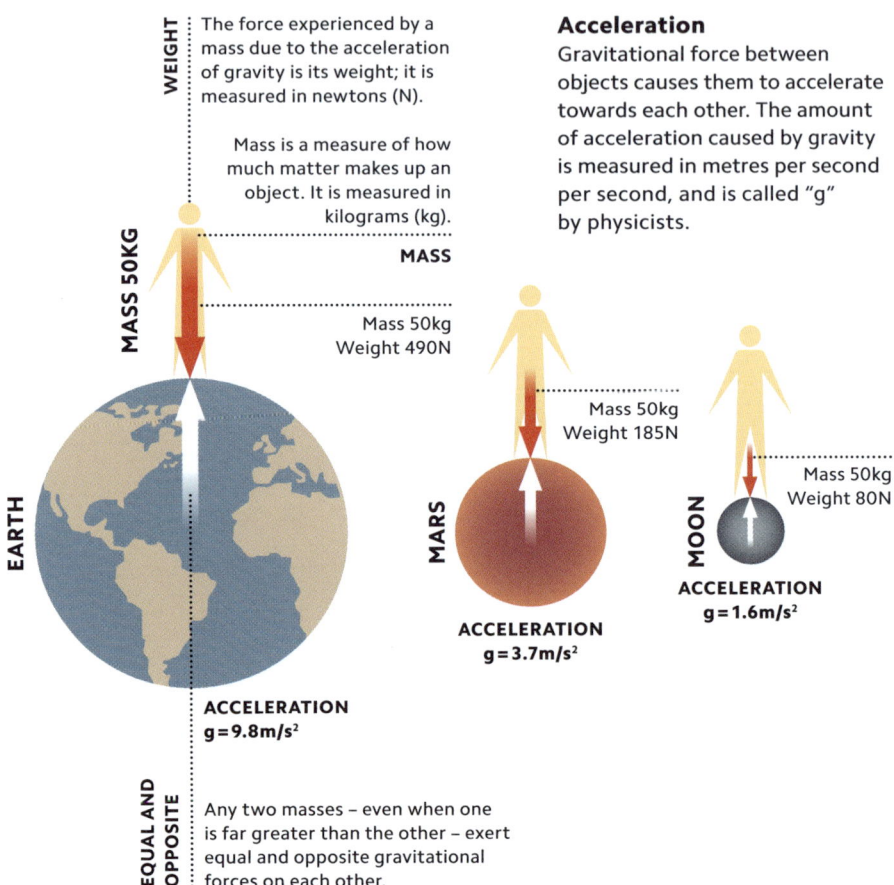

WEIGHT

The force experienced by a mass due to the acceleration of gravity is its weight; it is measured in newtons (N).

Mass is a measure of how much matter makes up an object. It is measured in kilograms (kg).

MASS

Mass 50kg
Weight 490N

MASS 50KG

EARTH

ACCELERATION
g = 9.8m/s²

EQUAL AND OPPOSITE

Any two masses – even when one is far greater than the other – exert equal and opposite gravitational forces on each other.

Acceleration

Gravitational force between objects causes them to accelerate towards each other. The amount of acceleration caused by gravity is measured in metres per second per second, and is called "g" by physicists.

Mass 50kg
Weight 185N

MARS

ACCELERATION
g = 3.7m/s²

Mass 50kg
Weight 80N

MOON

ACCELERATION
g = 1.6m/s²

FUNDAMENTAL ATTRACTION

Gravitation, or simply gravity, is a fundamental force (see pp.32–33). It is always attractive and affects all objects with mass – a measure of amount of matter. The greater the masses of two objects and the closer they are together, the greater their gravitational attraction. Modern physics does not consider gravitation as a force, but rather an effect caused by the curvature of spacetime (see p.136).

OPPOSING MOTION

Friction is a force between two surfaces in contact. It resists movement between the surfaces and is caused by the interaction between atoms and molecules at the microscopic scale. The size of the frictional force is proportional to the force pushing the two surfaces together, called the normal force. On a horizontal surface, the normal force is the weight of the object.

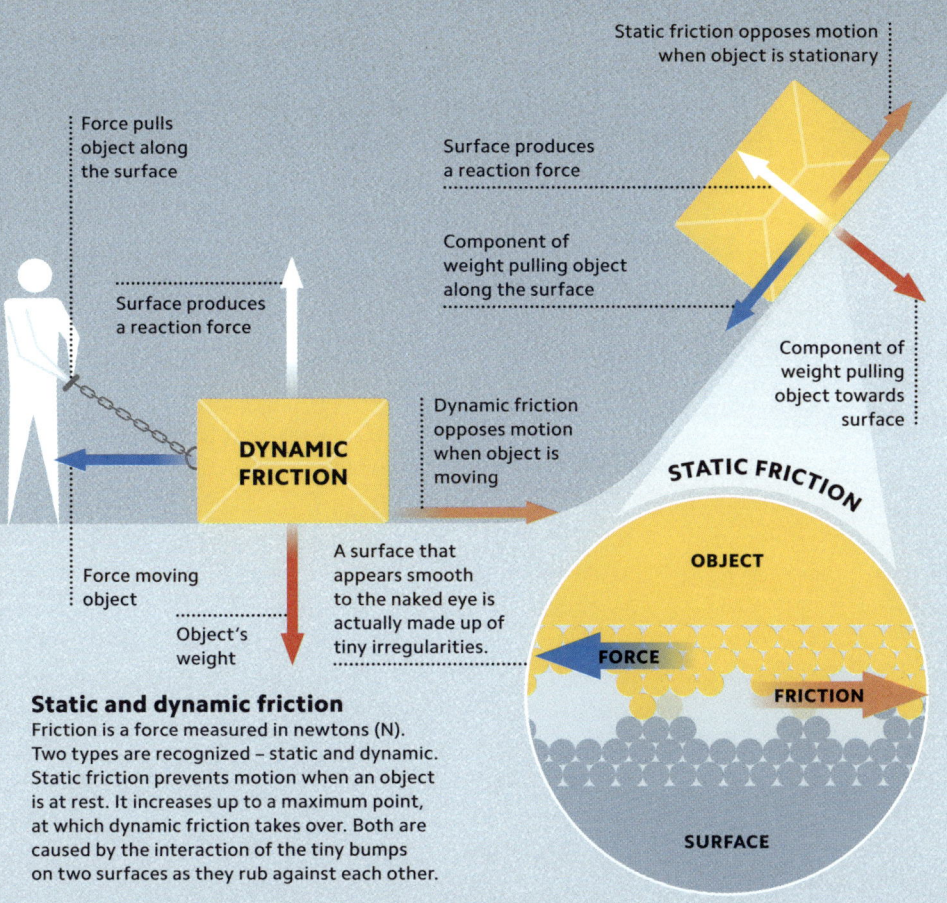

Static friction opposes motion when object is stationary

Force pulls object along the surface

Surface produces a reaction force

Surface produces a reaction force

Component of weight pulling object along the surface

Component of weight pulling object towards surface

DYNAMIC FRICTION

Dynamic friction opposes motion when object is moving

STATIC FRICTION

Force moving object

Object's weight

A surface that appears smooth to the naked eye is actually made up of tiny irregularities.

OBJECT

FORCE

FRICTION

SURFACE

Static and dynamic friction
Friction is a force measured in newtons (N). Two types are recognized – static and dynamic. Static friction prevents motion when an object is at rest. It increases up to a maximum point, at which dynamic friction takes over. Both are caused by the interaction of the tiny bumps on two surfaces as they rub against each other.

GAINING AN ADVANTAGE

Levers, wheels, axles, wedges, and pulleys are all simple machines –
devices that can change the direction and/or the size of an applied
force. The force put into a machine (the effort) can be the same as
the output force (the load); in such cases, the effort and load move
through identical distances. In other cases the effort and load are
different; the machine is then said to have a mechanical advantage
because either the output force or the distance moved is amplified.
However, since the machine produces a larger force moving over
a smaller distance (or vice versa), the amounts of
input work and output work are always equal.

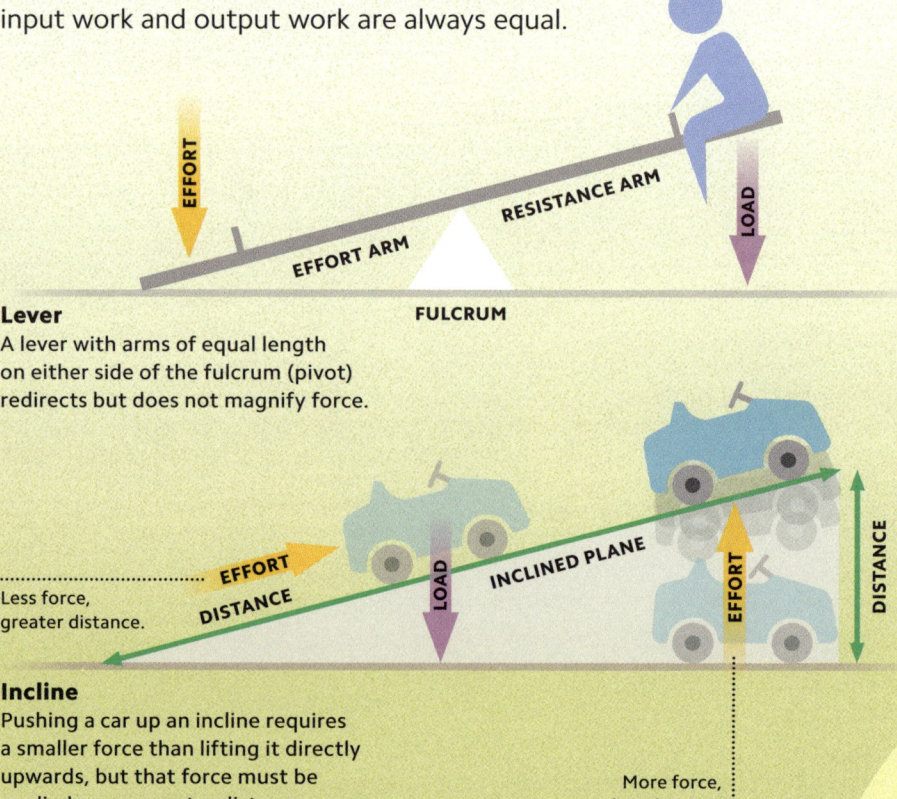

Lever
A lever with arms of equal length
on either side of the fulcrum (pivot)
redirects but does not magnify force.

Incline
Pushing a car up an incline requires
a smaller force than lifting it directly
upwards, but that force must be
applied over a greater distance.

Pulleys

Pulleys are simple machines made up of wheels and ropes or chains. By combining several pulleys, large mechanical advantages can be gained, though the input force must be applied over a longer distance.

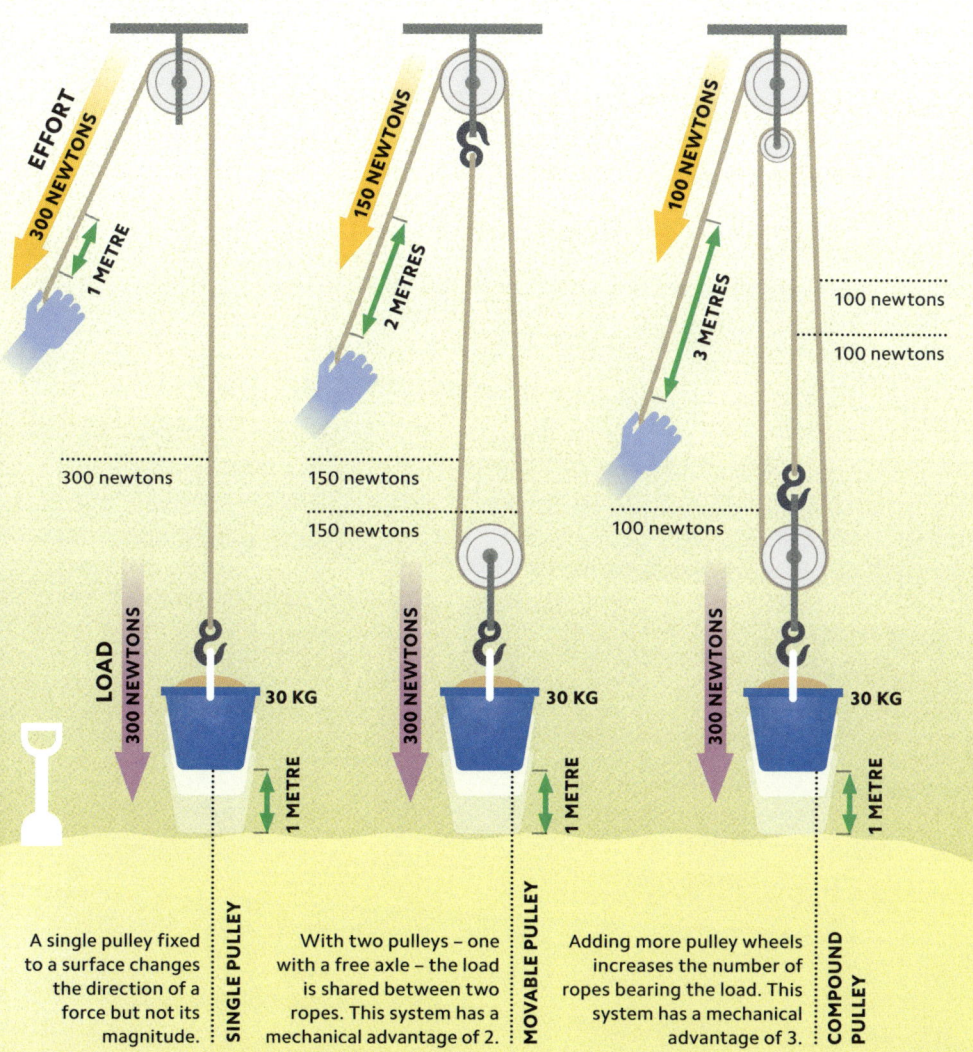

EFFORT
300 NEWTONS
1 METRE
300 newtons

150 NEWTONS
2 METRES
150 newtons
150 newtons

100 NEWTONS
3 METRES
100 newtons
100 newtons
100 newtons

LOAD
300 NEWTONS
30 KG
1 METRE

300 NEWTONS
30 KG
1 METRE

300 NEWTONS
30 KG
1 METRE

A single pulley fixed to a surface changes the direction of a force but not its magnitude.

SINGLE PULLEY

With two pulleys – one with a free axle – the load is shared between two ropes. This system has a mechanical advantage of 2.

MOVABLE PULLEY

Adding more pulley wheels increases the number of ropes bearing the load. This system has a mechanical advantage of 3.

COMPOUND PULLEY

UNDER PRESSURE

A fluid is a substance that can flow, such as a liquid or gas. The weight of a fluid creates pressure, which acts equally in all directions at a given depth. Pressure increases as depth increases. One consequence of this is buoyancy: the upward pressure on the bottom of a submerged object is greater than the downward pressure on the top, resulting in a net upward force, called upthrust. An object will float if the upthrust exceeds its weight. This occurs if the object's density is lower than that of the fluid.

Water spouts furthest from the bottom hole because the water pressure is greater at the bottom of the butt than at the top.

WATER SPOUTS

ATMOSPHERIC PRESSURE The atmosphere is a fluid and exerts its own pressure, which increases with depth.

FLOAT A hollow ball floats because its weight is lower than the upthrust by fluid pressure.

UPTHRUST Pressure is greater on the bottom of the ball than on top.

PRESSURE CHANGE Pressure depends on the density of the fluid and its depth.

WATER BUTT

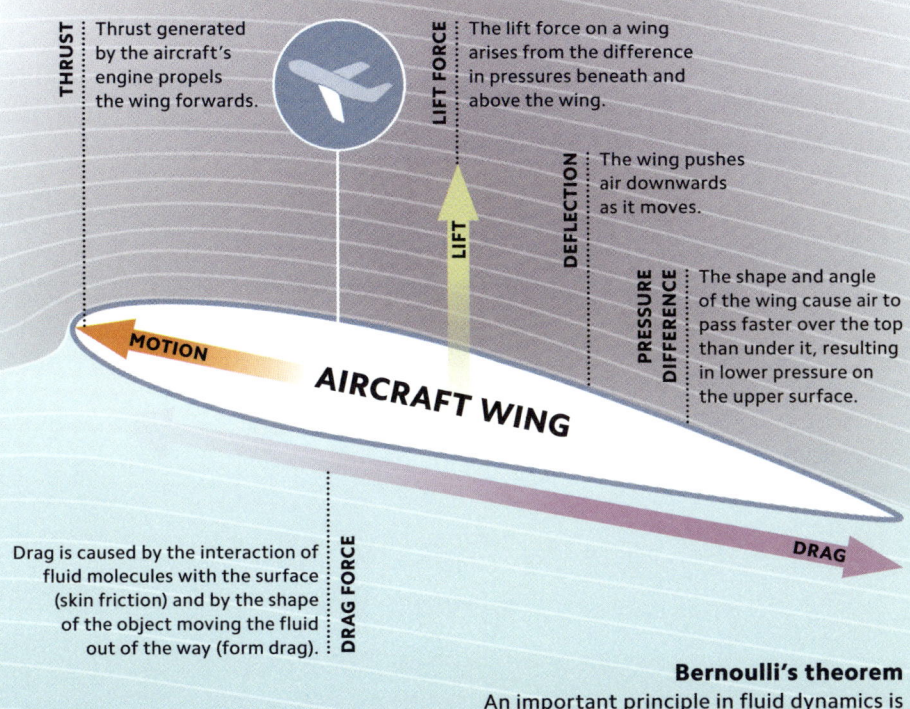

THRUST Thrust generated by the aircraft's engine propels the wing forwards.

LIFT FORCE The lift force on a wing arises from the difference in pressures beneath and above the wing.

LIFT

DEFLECTION The wing pushes air downwards as it moves.

PRESSURE DIFFERENCE The shape and angle of the wing cause air to pass faster over the top than under it, resulting in lower pressure on the upper surface.

MOTION

AIRCRAFT WING

DRAG

DRAG FORCE Drag is caused by the interaction of fluid molecules with the surface (skin friction) and by the shape of the object moving the fluid out of the way (form drag).

Bernoulli's theorem
An important principle in fluid dynamics is Bernoulli's theorem, which states that as the speed of a fluid past a surface increases, the pressure it exerts on the surface decreases.

FLOWING FLUIDS

Fluid dynamics is the study of how fluids (liquids and gases) move when forces are applied to them. It includes aerodynamics (with air as the fluid) and hydrodynamics (with water as the fluid). Two forces that arise in fluids are drag and lift. Like friction, drag acts in the direction against motion. Its size depends upon the density, viscosity, and compressibility of the fluid, the relative speed of object and fluid, and the shape and size of the object. Lift occurs where the pressure exerted by a fluid is greater at some points on an object than at others.

ENERG

One of the most important concepts in physics is that of energy – the capacity to elicit change or, more technically, to do work (see pp.56–57). Energy changes can be calculated in different situations such as when a force moves an object, when a hot object heats a colder one, or when chemical substances react. A key aspect of energy is that it cannot be created or destroyed, so the total amount of energy at the end of a process is the same as at the start, but it will have been redistributed between the interacting bodies. This law of conservation allows scientists to calculate, for example, how an object will move or the extent to which two substances will react together.

PUTTING FORCES TO WORK

A force acting on an object can increase the object's energy. It can make the object go faster (increasing its kinetic energy), lift it to a higher position (increasing its gravitational potential energy), or stretch it (increasing its store of elastic energy). A force is said to do work when it transfers energy to an object, and the increase in the object's energy is equal to the work done on it. Both work and energy are measured in joules (J).

Work

The bigger the force acting on an object and the further it pushes or pulls on the object, the greater the work done by the force and hence the greater the energy transferred to the object.

NO WORK
No work is done on this box if the box does not move, even though a force is applied.

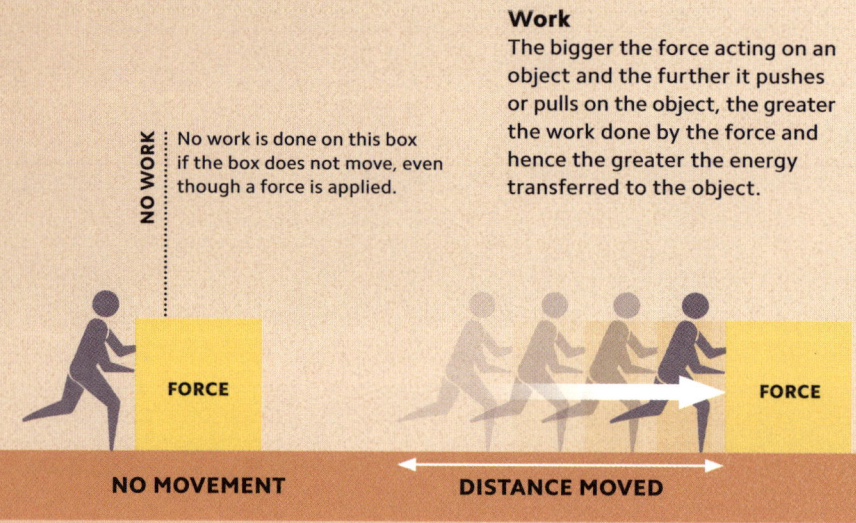

FORCE

NO MOVEMENT

FORCE

DISTANCE MOVED

Kinetic energy

Kinetic energy is the energy possessed by an object by virtue of its motion. It depends on the object's mass and the square of its velocity – so even small objects can carry lots of energy if they move fast.

SPEED 5KPH

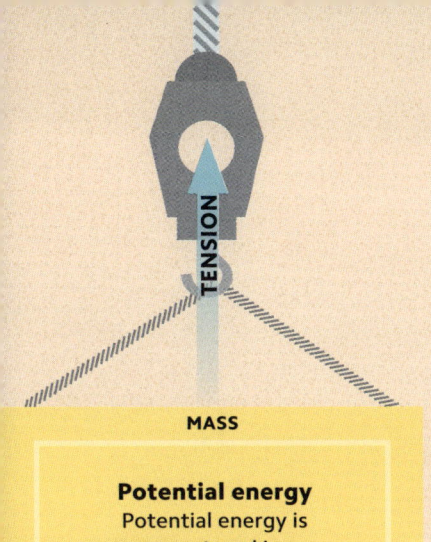

TENSION

MASS

Potential energy
Potential energy is energy stored in an object by virtue of its position or shape. Lifting an object against the force of gravity increases its potential energy, as does compressing a spring to change its shape.

GRAVITY

Energy conversion
Potential energy is converted into kinetic energy when the object is dropped.

LIFTING

The potential energy of a raised object depends on its mass, the height it has been raised, and the strength of gravity.

HEIGHT

SPEED 1,500KPH **10G**

1,000KG

ENERGY EQUIVALENCE
A car with a mass of 1,000kg (2,200lb) moving at 5kph (3mph) has about the same kinetic energy as a 10g bullet travelling at 1,500kph (930mph).

Power outputs

The amounts of power produced or used by different processes vary enormously. The output of the Sun, for example, is around 4×10^{26} watts, while a wristwatch consumes less than 10 microwatts.

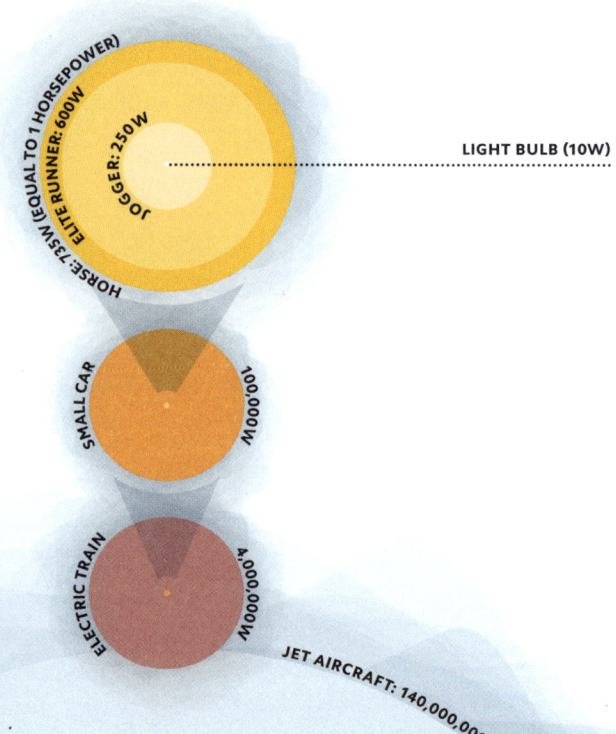

JOGGER: 250W

HORSE: 735W (EQUAL TO 1 HORSEPOWER) ELITE RUNNER: 600W

LIGHT BULB (10W)

SMALL CAR

100,000W

ELECTRIC TRAIN

4,000,000W

JET AIRCRAFT: 140,000,000W

ENERGY BY NUMBERS

Energy can be defined as the capacity to do work – the ability to move an object over a distance (see p.56). Several units are used to quantify amounts of energy: the amount of energy in food is measured in calories; the energy produced by an explosion may be expressed in terms of tonnes of TNT; and energy released when fuels are burned may be given in thermal units. The standard scientific unit of energy, however, is the joule (J). Power is a different concept; it is the rate of transfer of energy, or the energy transferred per second. Its unit is the watt (W), equal to one joule per second.

WHERE DOES ENERGY GO?

When an apple falls to the ground, an electric current causes a bulb to glow, or a chemical reaction makes a firework explode, the total amount of energy in the world remains constant. The law of conservation of energy states that energy cannot be created or destroyed, only converted from one form to another. In practice, energy tends to dissipate, becoming more spread out and so less useful (see pp.66–67).

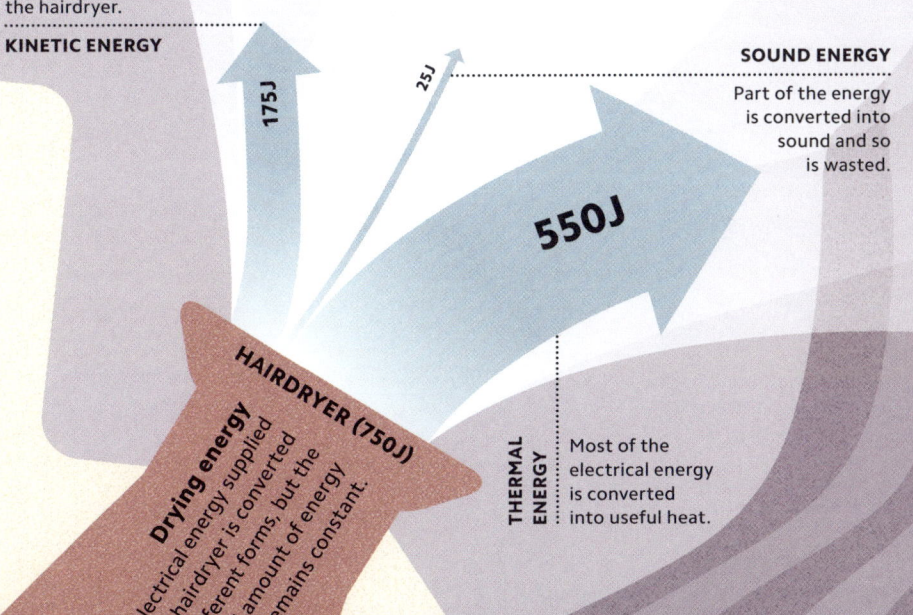

A portion of the electrical energy drives a fan that propels air from the hairdryer.

KINETIC ENERGY

175J

25J

SOUND ENERGY

Part of the energy is converted into sound and so is wasted.

550J

HAIRDRYER (750J)

Drying energy
The electrical energy supplied to a hairdryer is converted into different forms, but the total amount of energy remains constant.

THERMAL ENERGY

Most of the electrical energy is converted into useful heat.

MOLECULES ON THE MOVE

The atoms and molecules of a system are in random motion: they may vibrate, rotate, or move in any direction, or all three. The energy of this random motion increases with temperature and is known as thermal energy. It can be transferred from one system to another in the form of heat. It always moves from a hotter to a cooler place – the greater the temperature difference, the greater the rate at which energy will be transferred.

Particle energies

Temperature is a measure of the average kinetic energy of particles within a system; it can be measured on scales such as the Celsius and Fahrenheit scales. The graph below shows the range of kinetic energies of the particles in a substance at a given time.

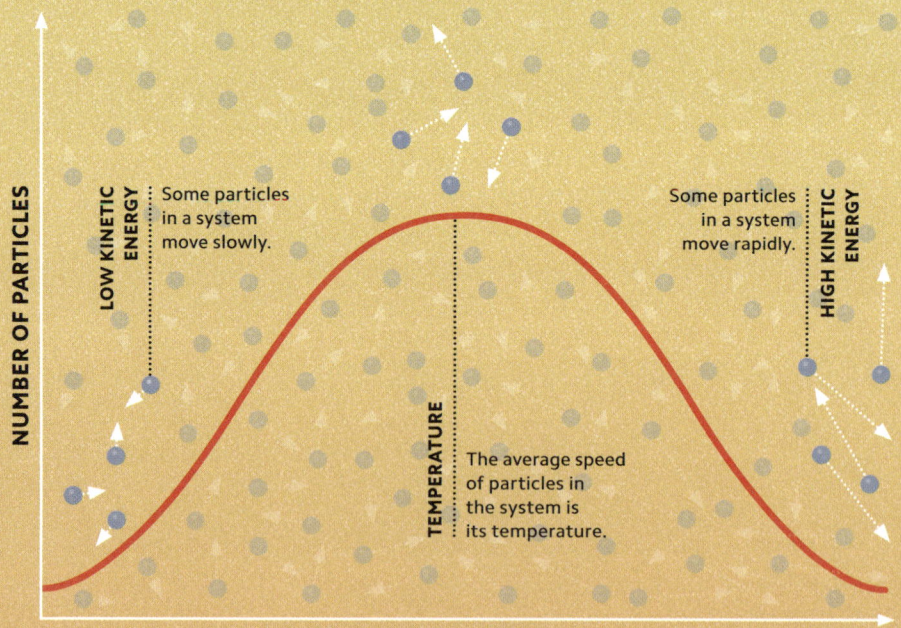

NUMBER OF PARTICLES

LOW KINETIC ENERGY

Some particles in a system move slowly.

HIGH KINETIC ENERGY

Some particles in a system move rapidly.

TEMPERATURE

The average speed of particles in the system is its temperature.

KINETIC ENERGY

Temperature scales

Temperature is measured on the Celsius and Fahrenheit scales, which have reference points based on the boiling and melting points of water and other substances, or on the Kelvin scale, which starts from the lowest temperature possible.

BOILING

Water molecules move rapidly and can move past one another.

FREEZING

Water molecules are locked in position but can still vibrate.

ABSOLUTE ZERO

Particles cease all movement. Absolute zero is the lowest temperature possible, but cannot be achieved in practice.

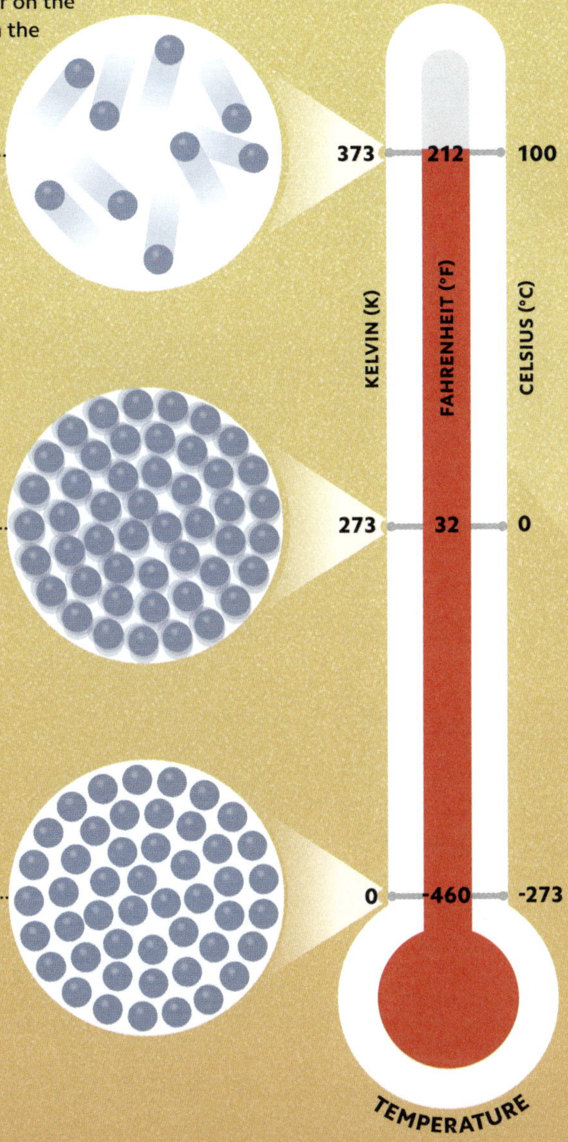

KELVIN (K)

FAHRENHEIT (°F)

CELSIUS (°C)

373   212   100

273   32   0

0   -460   -273

TEMPERATURE

SPREADING ENERGY

Heating is the term commonly used to describe the transfer of energy from a hotter object to a colder one. There are three ways that this transfer can happen. All objects above a temperature of absolute zero (see p.61) transfer heat by radiation – electromagnetic waves that transmit energy between objects that are not in contact. Conduction transfers energy between objects that are in contact. Convection is the transfer of energy through a fluid.

Radiation

Electromagnetic radiation – such as light, infrared, and ultraviolet radiation – can transfer energy either through a medium or through a vacuum, as happens for example when energy from the Sun heats Earth.

NIGHT LOSSES

At night, the side of Earth that faces into the darkness of space receives no radiation, so cools as it emits infrared.

LOSING ENERGY

Earth constantly emits infrared radiation into the coldness of space.

SOLAR SOURCE

With a surface temperature of 4,500°C (8,100°F), the Sun is a strong emitter of electromagnetic radiation.

ATMOSPHERIC BLANKET

The atmosphere absorbs and re-radiates infrared from Earth's surface, increasing temperatures. This is known as the greenhouse effect.

INFRARED HEATING

The Sun emits a wide spectrum of radiation, but infrared wavelengths are most important in warming Earth.

Conduction

When a solid is heated, its constituent particles vibrate rapidly. When they collide with their neighbours, they pass on some of their kinetic energy.

METAL PROPERTIES

Metals are excellent conductors of heat. Kinetic energy is transferred not only by the vibration of atoms but also by the movement of free electrons (see p.17).

VIBRATIONS

Particles at the heated end of this metal bar vibrate more rapidly than those at the cool end. Vibrational energy spreads through the metal. The molecules vibrate but do not change their position in the bar.

HEAT SOURCE

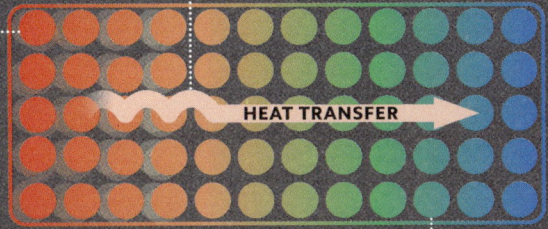

HEAT TRANSFER

THERMAL GRADIENT

The difference in temperature between the two ends of the bar drives the transfer of energy. Once the entire bar is hot, the rate of energy transfer decreases.

Convection

Convection occurs in liquids and gases. When heated, pockets of fluid become less dense and so rise upwards through the body of the fluid. Cooler fluid sinks before being heated itself.

CONVECTION CURRENT

Warm water rises above a heater; colder water flows in to replace it.

FASTER MOLECULES

As the water is heated, its molecules move faster and push each other further apart.

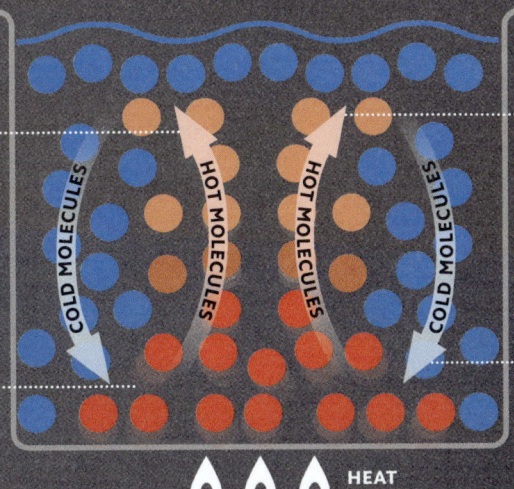

COLD MOLECULES

HOT MOLECULES

HOT MOLECULES

COLD MOLECULES

HEAT SOURCE

CHANGING DENSITY

The density of the water decreases as its molecules move faster. The less dense water floats upwards.

SINKING

Cooling particles have less energy and higher density, so sink back into the body of the liquid.

Specific heat capacity

The energy required to raise the temperature of 1kg of a substance by 1°C is called its specific heat capacity (SHC). Liquids such as water have a high SHC and are suited for use as coolants in engines because they can carry away excess heat without becoming very hot or boiling.

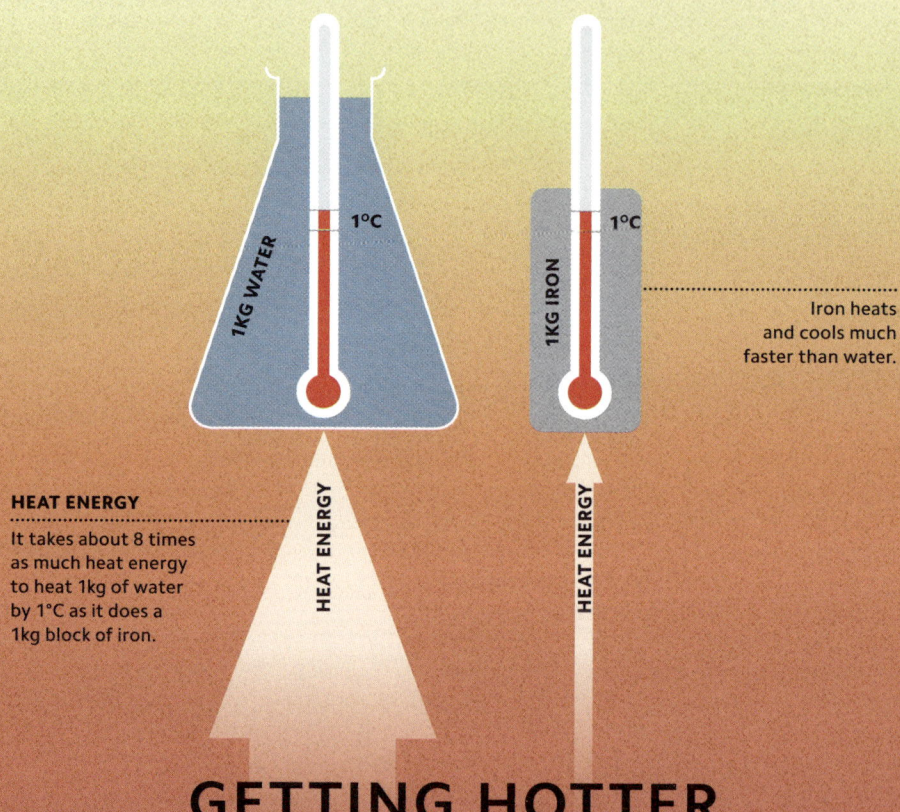

1°C

1KG WATER

1°C

1KG IRON

Iron heats and cools much faster than water.

HEAT ENERGY

It takes about 8 times as much heat energy to heat 1kg of water by 1°C as it does a 1kg block of iron.

HEAT ENERGY

HEAT ENERGY

GETTING HOTTER

When matter is heated, its temperature rises (see pp.60–61) and it may change state (from solid to liquid, or liquid to gas). Materials differ in the rate at which they respond to heat input. It takes much more heat to raise the temperature of a substance that has strong bonds between its molecules, such as water, than one with bonds that allow its particles to move more freely, such as a metal.

HIDDEN HEAT

When heat is applied to a solid, its temperature rises until it eventually melts to form a liquid and then again until it vaporizes, forming a gas. At each change of state (or phase), extra energy must be supplied to bring about the transition from one state to the other; this is called latent (or hidden) heat. Input of this energy does not cause a temperature change but is required to overcome the intermolecular forces that hold particles together in a solid, and again to let the molecules escape as a gas.

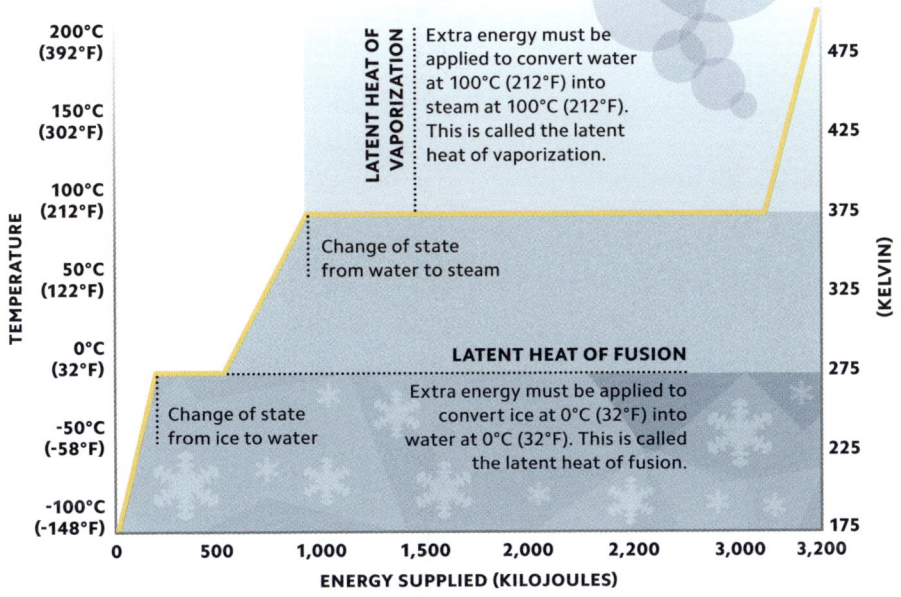

LATENT HEAT OF VAPORIZATION

Extra energy must be applied to convert water at 100°C (212°F) into steam at 100°C (212°F). This is called the latent heat of vaporization.

Change of state from water to steam

LATENT HEAT OF FUSION

Extra energy must be applied to convert ice at 0°C (32°F) into water at 0°C (32°F). This is called the latent heat of fusion.

Change of state from ice to water

TEMPERATURE

200°C (392°F)	475
150°C (302°F)	425
100°C (212°F)	375
50°C (122°F)	325
0°C (32°F)	275
-50°C (-58°F)	225
-100°C (-148°F)	175

(KELVIN)

ENERGY SUPPLIED (KILOJOULES)

0   500   1,000   1,500   2,000   2,200   3,000   3,200

BECOMING RANDOM

Entropy is a measure of the randomness of the particles of a system. The more random the arrangement of the particles, the more the energy of the molecules in the system is dispersed, and the greater the entropy. The level of entropy determines how much work (such as mechanical work) can be done by a system and predicts the direction of a process (such as a chemical reaction).

Mixing gases
Here are two chambers, one filled with hot gas molecules (that move rapidly) and another with cold gas molecules (that move slowly).

Entropy and time
The laws of probability dictate that entropy increases with time and that – eventually – all energy and matter will be evenly dispersed in the Universe. These processes all move one way in time, so giving time a "direction".

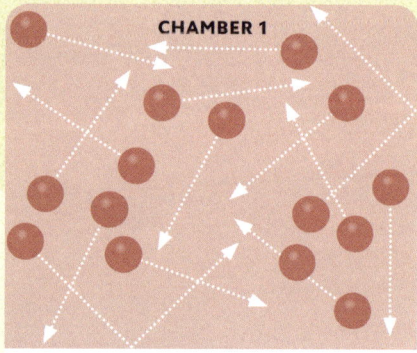

CHAMBER 1

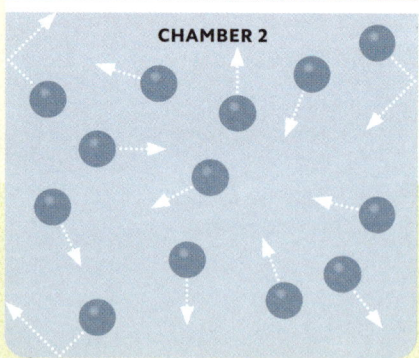

CHAMBER 2

LOW ENTROPY

The temperature difference between the "hot" gas molecules in chamber 1 and the "cold" gas molecules in chamber 2 can be used to transfer energy and so do work.

> "Disorder increases with time because we measure time in the direction in which disorder increases."
> Stephen Hawking

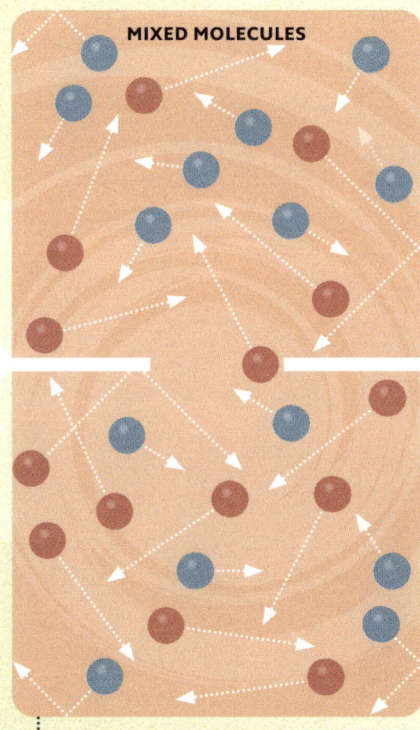

MIXED MOLECULES

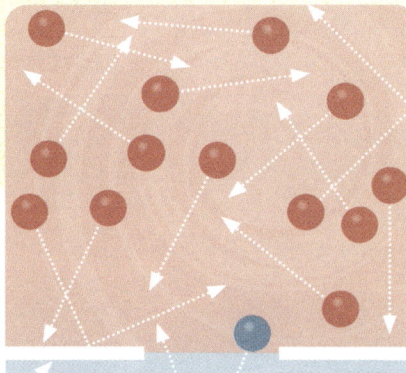

HIGH ENTROPY

Joining the chambers mixes hot and cool molecules. There is no longer a temperature difference between the chambers because the system is more disordered. It cannot do any useful work.

IMPROBABLE OUTCOMES

The hot molecules could, in theory, all move back into one chamber and the cold molecules into the other. This would create a system with low entropy. In practice, however, this is highly improbable; entropy always increases.

THE LAWS OF ENERGY

Thermodynamics is the science of energy. It has four fundamental laws, numbered from zero to three. (The zeroth law was recognized later than the other three, although it precedes them logically.) There are different ways in which the laws are expressed, but they are all concerned with the way energy is transferred from one object to another, how energy tends to spread out, and how entropy increases.

Zeroth law

If there is no net flow of energy between two objects, they are said to be in thermal equilibrium. They are at the same temperature.

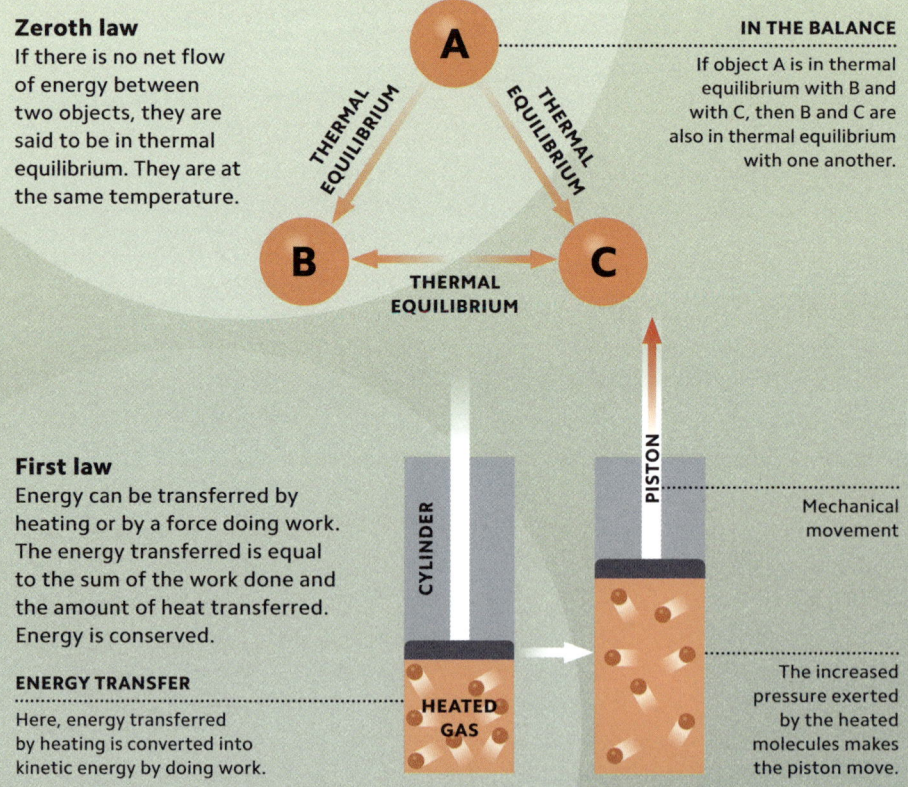

IN THE BALANCE

If object A is in thermal equilibrium with B and with C, then B and C are also in thermal equilibrium with one another.

First law

Energy can be transferred by heating or by a force doing work. The energy transferred is equal to the sum of the work done and the amount of heat transferred. Energy is conserved.

ENERGY TRANSFER

Here, energy transferred by heating is converted into kinetic energy by doing work.

Mechanical movement

The increased pressure exerted by the heated molecules makes the piston move.

Second law

Energy tends to dissipate – to spread out into the environment. This happens both when a force does work and when heat transfer results from a temperature difference. As a result, the system becomes more disordered – its entropy increases (see pp.66–67).

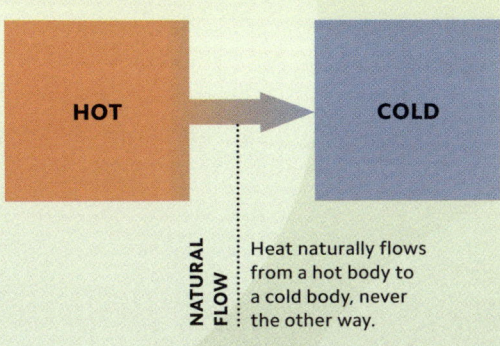

HOT

COLD

NATURAL FLOW

Heat naturally flows from a hot body to a cold body, never the other way.

Third law

To cool an object, energy must be removed from it. This becomes increasingly difficult as the temperature decreases; it is impossible to attain the temperature of absolute zero (see page 64).

The Large Hadron Collider operates at a temperature of just 1.9K.

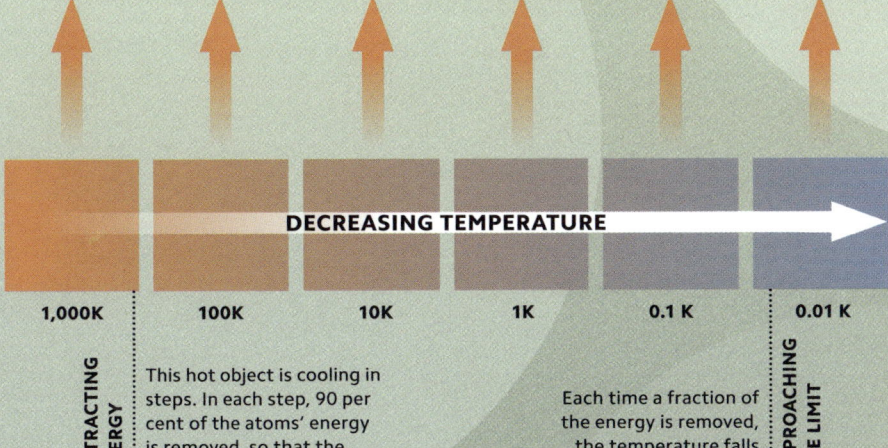

90%   90%   90%   90%   90%   90%

DECREASING TEMPERATURE

1,000K   100K   10K   1K   0.1 K   0.01 K

EXTRACTING ENERGY

This hot object is cooling in steps. In each step, 90 per cent of the atoms' energy is removed, so that the temperature divides by 10.

Each time a fraction of the energy is removed, the temperature falls but it never reaches 0K.

APPROACHING THE LIMIT

ELECTR

AND

MAGNET

CITY

S M

Electric charge is an intrinsic property of elementary particles, such as protons and electrons. Electric charges create electric fields, which we can feel as static electricity. Moving charges, such as electrons in a wire, form an electric current, generating a magnetic field in the process. Electromagnetism explains how electric currents create magnetic fields and vice versa. This principle is crucial in motors, which convert electrical energy into mechanical motion, and generators, which do the reverse. Electromagnetic waves, such as light, are formed by oscillating electric and magnetic fields travelling through space at the speed of light.

The strength and direction of a field are represented using field lines.

The closer the lines are spaced, the stronger the electric field.

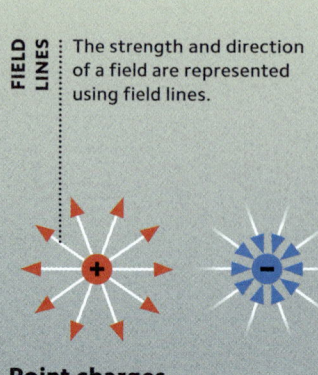

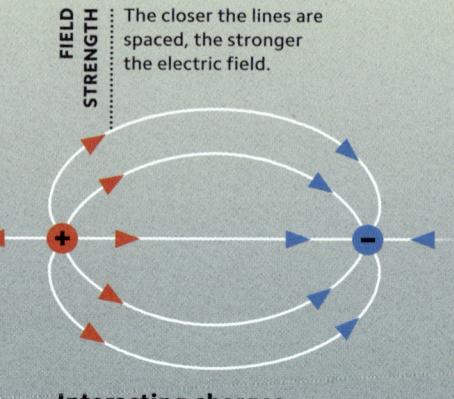

Point charges
A charged particle has a symmetrical electric field. The field is a vector quantity (see p.38) – one that has direction as well as size.

Interacting charges
Two opposite charges produce the field shown above; field lines are directed from positive to negative.

FIELDS OF FORCE

Electric charge is a fundamental property of subatomic particles). There are two types of charge – positive and negative. Two (or more) charged particles exert forces on one other without having to touch; like charges repel, while unlike charges attract. Any region in which charged particles experience such forces is called an electric field. The strength of the field, and so the size of the forces, depends upon the amount of charge present and the distance between charged particles. Under certain circumstances, such as when objects rub together, electric charges can accumulate on their surfaces, a phenomenon called static electricity.

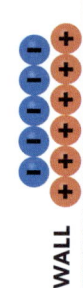

The balloon's surface has a negative charge that repels negative charge in the wall.

BALLOON

WALL

The wall's surface is left with a positive charge and so attracts the balloon.

WALL

Static electricity
Rubbing a balloon on a woollen jumper transfers electrons onto the balloon, giving it a negative charge.

MOVING CHARGES

An electric current is a flow of charged particles. Moving protons or ions can make up an electric current, but the most familiar charge carriers are electrons (see p.73), which flow through wires in most circuits. A potential difference, measured in volts (V), is needed to push electrons around a circuit. This is provided by a battery or the mains supply. The greater the voltage the greater the potential energy of the electrons. The magnitude of the current – the amount of electric charge passing one point in a second – is measured in amps (A).

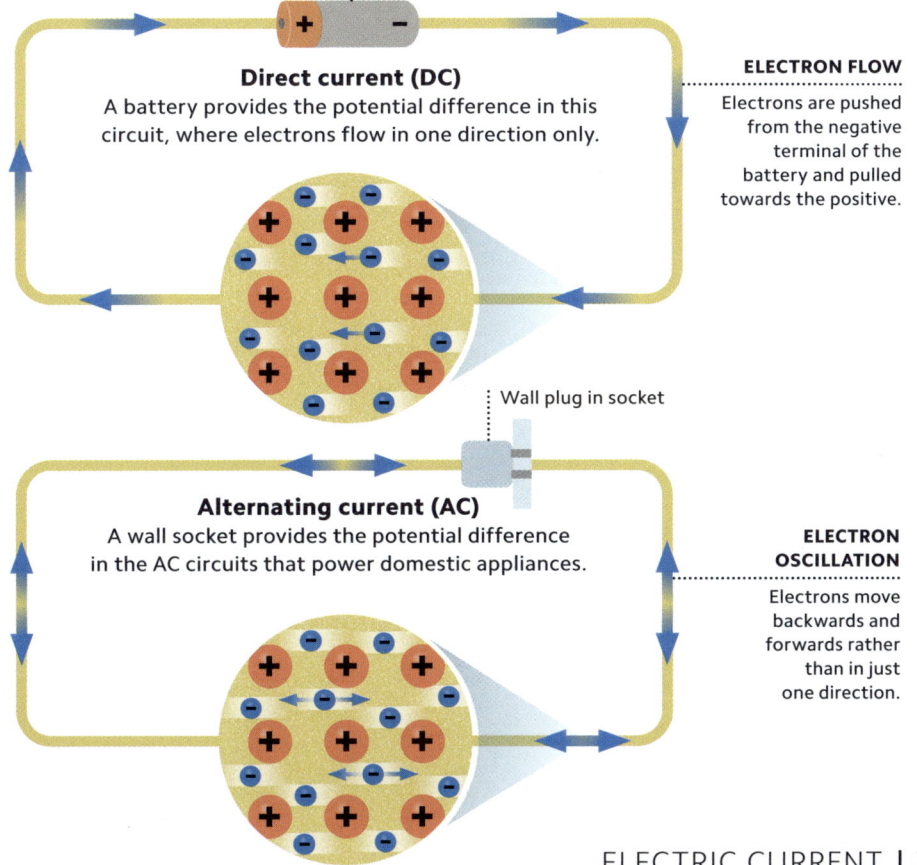

Battery

Direct current (DC)
A battery provides the potential difference in this circuit, where electrons flow in one direction only.

ELECTRON FLOW
Electrons are pushed from the negative terminal of the battery and pulled towards the positive.

Wall plug in socket

Alternating current (AC)
A wall socket provides the potential difference in the AC circuits that power domestic appliances.

ELECTRON OSCILLATION
Electrons move backwards and forwards rather than in just one direction.

ELECTRIC CURRENT | 73

PORTABLE POWER

An electric cell is a device that stores chemical energy. In a circuit, the potential difference (see p.73) between the two electrodes powers devices such as motors, lighting, and consumer electronics. A battery is made up of several cells connected to provide a higher voltage. Single-use cells and batteries contain compounds that are used up once their energy is depleted. Rechargeable batteries, such as lithium-ion cells, can be replenished by connecting to an external source of electrical power.

CAP

ANODE (-)
SEPARATOR
CATHODE (+)

Stored energy
The battery of an electric vehicle (EV) is typically made up of several thousand cylindrical cells, each with a voltage of 3–4V. They are connected together to produce a voltage of around 400V that powers the vehicle's motors.

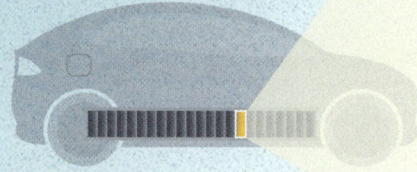

LITHIUM-ION BATTERY PACK IN EV

Cell structure
A lithium-ion cell is made of a metal oxide cathode separated from a graphite anode by a thin sheet that allows the passage of ions but not of electrons. This sheet is moistened with an electrolyte solution.

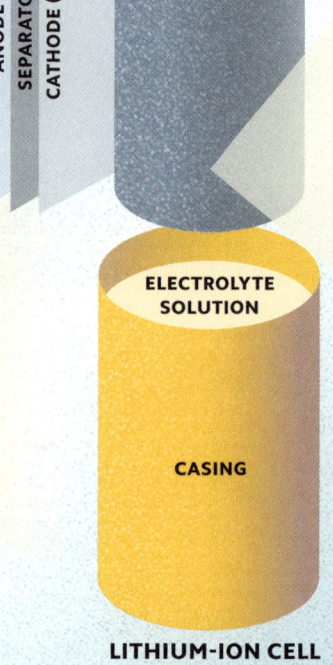

ELECTROLYTE SOLUTION

CASING

LITHIUM-ION CELL

FLAT BATTERY

ELECTROLYTE
This fluid is able to conduct current within the battery.

LITHIUM ATOMS
When the battery is flat, lithium atoms are attached to the metal oxide of the positive electrode.

POSITIVE ELECTRODE
This is made of lithium atoms and oxygen atoms bonded together.

NEGATIVE ELECTRODE
This is made of graphite, a form of the element carbon.

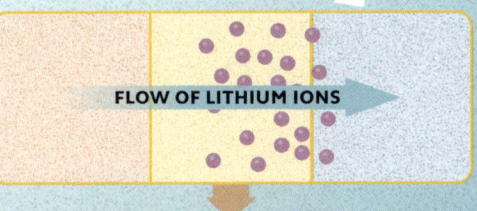

Charge and discharge
Lithium-ion cells work by moving ions from one electrode to the other during charging and discharging cycles.

CHARGING UP
The power supply pushes electrons from the positive terminal to the negative. Their negative charge attracts a flow of positive lithium ions.

FLOW OF LITHIUM IONS

FULLY CHARGED
The lithium atoms now reside at the negative graphite electrode.

PROVIDING POWER
The flow of electrons from the negative electrode drives the EV's electric motor (see opposite).

MOTOR

DISCHARGING
When power is needed, electrons leave the lithium atoms and flow through the motor. The positively charged lithium ions flow back through the electrolyte.

FLOW OF LITHIUM IONS

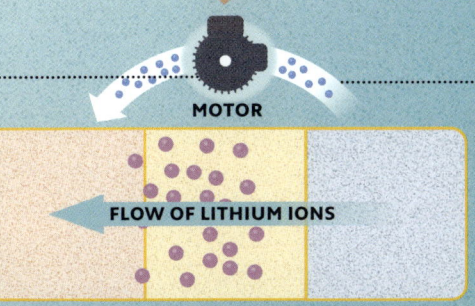

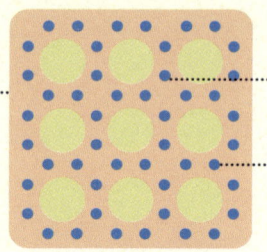

PURE SILICON

Pure silicon is a crystalline solid. A silicon atom has four electrons, each of which forms a covalent bond with an adjacent silicon atom.

Each of a silicon atom's four outer electrons bonds with another silicon atom.

COVALENT BONDS

CONDUCTIVITY

Electrons are bound only loosely, giving pure silicon moderate conductivity.

SILICON CRYSTAL

CURRENT CONTROL

The flow of charged particles, such as electrons or ions, is an electric current. Conductors, such as metals, have plenty of free electrons in their structure, while insulators have few particles available to carry charge. A semiconductor, such as silicon, is between the two; it can be made to conduct by boosting the energy level of its electrons by exposing it to light or heat. Its properties can also be changed by doping – adding atoms of other specific elements to its structure. This creates a material that has an excess or deficit of electrons in its structure – an n-type or p-type semiconductor.

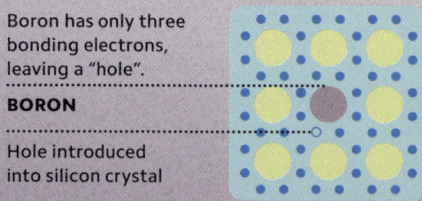

Boron has only three bonding electrons, leaving a "hole".

BORON

Hole introduced into silicon crystal

P-type silicon
Doping with boron atoms introduces holes into the crystal. The material can now conduct electricity because electrons can move into the holes.

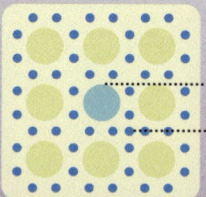

A phosphorus atom has five bonding electrons, adding an extra, mobile electron.

PHOSPHORUS

Extra electron introduced into silicon crystal

N-type silicon
Doping with phosphorus atoms introduces more electrons into the crystal; these electrons carry charge, increasing conductivity.

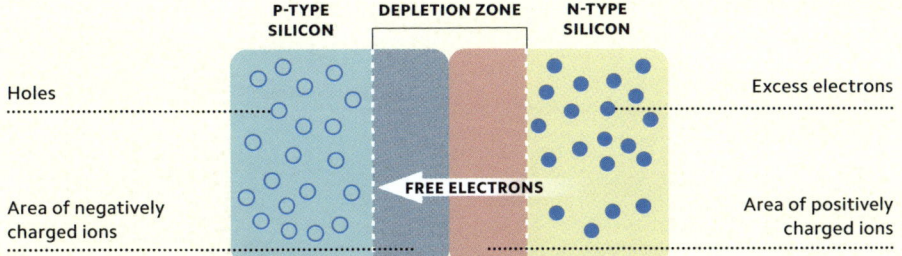

P-N junctions

When p-type and n-type silicon are placed into contact, some excess electrons from the n-type silicon cross over to fill the holes in the adjacent p-type material. This forms a depletion zone that is almost non-conductive because it contains no mobile charges.

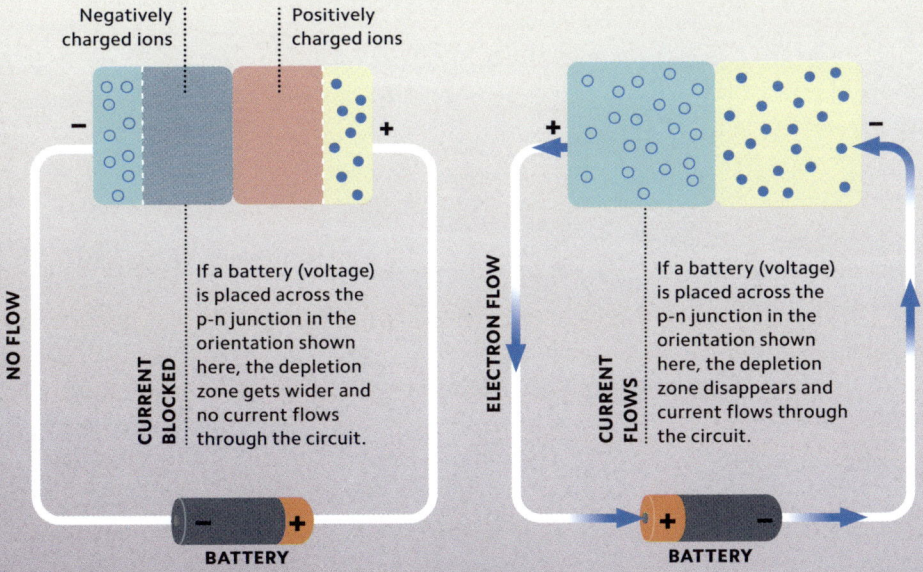

If a battery (voltage) is placed across the p-n junction in the orientation shown here, the depletion zone gets wider and no current flows through the circuit.

If a battery (voltage) is placed across the p-n junction in the orientation shown here, the depletion zone disappears and current flows through the circuit.

Diodes

A diode is a simple semiconductor device made up of p-type and n-type material. It allows current to flow through it in just one direction. Diodes are used to convert AC to DC current (see p.73).

(see p.73)

ELECTRIC LOGIC

Transistors are electronic components that rely on semiconductors. They are most often used as electronic switches. A small voltage applied to one of the transistor's three terminals creates an electric field that lets current flow between the other two terminals. Transistors can be individual components, but are also present in their billions on microchips, where they represent digital 1s and 0s as they turn on and off.

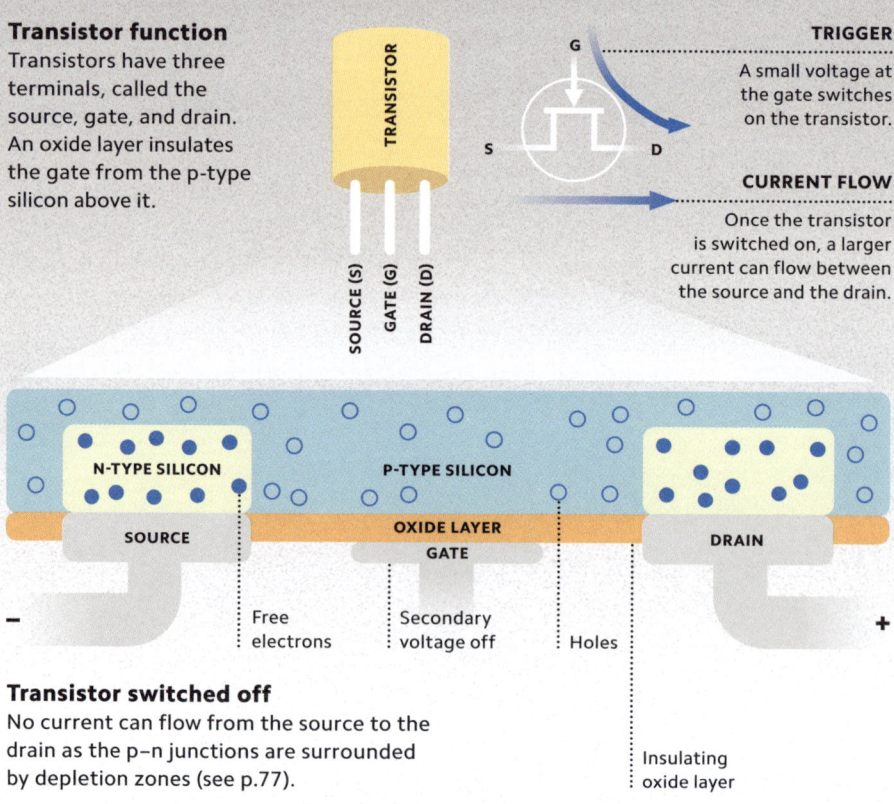

Transistor function
Transistors have three terminals, called the source, gate, and drain. An oxide layer insulates the gate from the p-type silicon above it.

TRANSISTOR

SOURCE (S)   GATE (G)   DRAIN (D)

G

S   D

TRIGGER
A small voltage at the gate switches on the transistor.

CURRENT FLOW
Once the transistor is switched on, a larger current can flow between the source and the drain.

N-TYPE SILICON   P-TYPE SILICON

SOURCE   OXIDE LAYER   DRAIN
GATE

−   +

Free electrons | Secondary voltage off | Holes

Transistor switched off
No current can flow from the source to the drain as the p–n junctions are surrounded by depletion zones (see p.77).

Insulating oxide layer

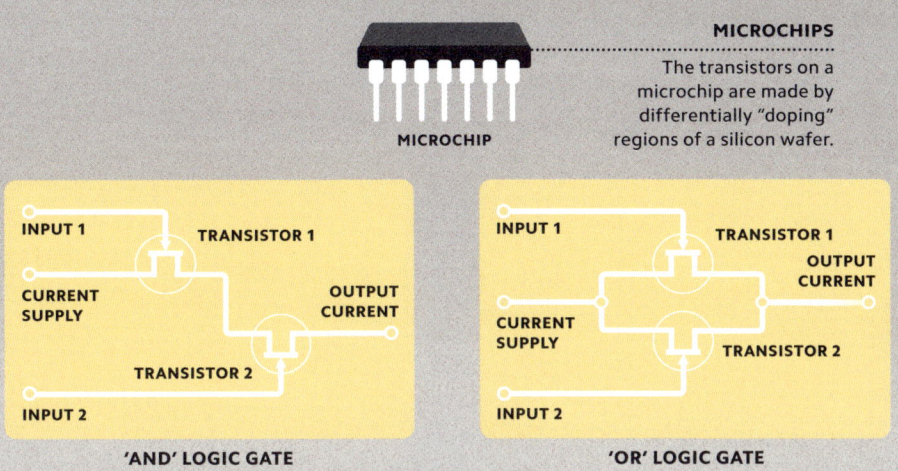

MICROCHIPS
The transistors on a microchip are made by differentially "doping" regions of a silicon wafer.

MICROCHIP

INPUT 1

TRANSISTOR 1

CURRENT SUPPLY

OUTPUT CURRENT

TRANSISTOR 2

INPUT 2

'AND' LOGIC GATE

INPUT 1

TRANSISTOR 1

OUTPUT CURRENT

CURRENT SUPPLY

TRANSISTOR 2

INPUT 2

'OR' LOGIC GATE

Logic switches

The transistors on a microchip can be arranged in configurations called gates to carry out logical operations. In the AND gate, both transistors must be switched on to obtain an output current; in an OR gate, either one or the other transistor must be on to obtain an output current.

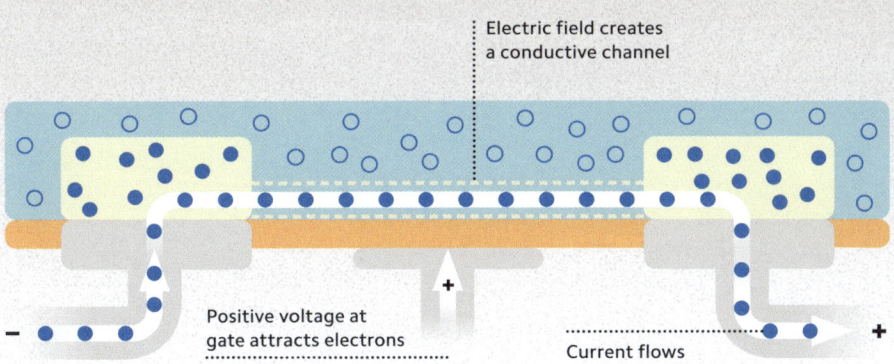

Electric field creates a conductive channel

Positive voltage at gate attracts electrons

Current flows

Transistor switched on

A positive voltage applied to the gate creates an electric field that attracts electrons, making a channel through which current can flow.

POLAR FORCES

Any moving electric charge creates a magnetic field. Every electron in an atom spins, resulting in a tiny magnetic field. Usually, the billions of electrons in a object spin in many different orientations, resulting in no overall magnetic field. However, in certain materials, such as iron, these tiny atomic magnets can be lined up to give a bar magnet a permanent magnetic field.

Domains

Iron is made up of small regions called domains in which all the atomic magnets are aligned. When exposed to a strong magnetic field, the domains can be brought into alignment, giving an iron bar an overall magnetic field.

MAGNETIZED Domains are aligned in a magnetized iron bar.

UNMAGNETIZED Domains are randomly oriented in an unmagnetized iron bar.

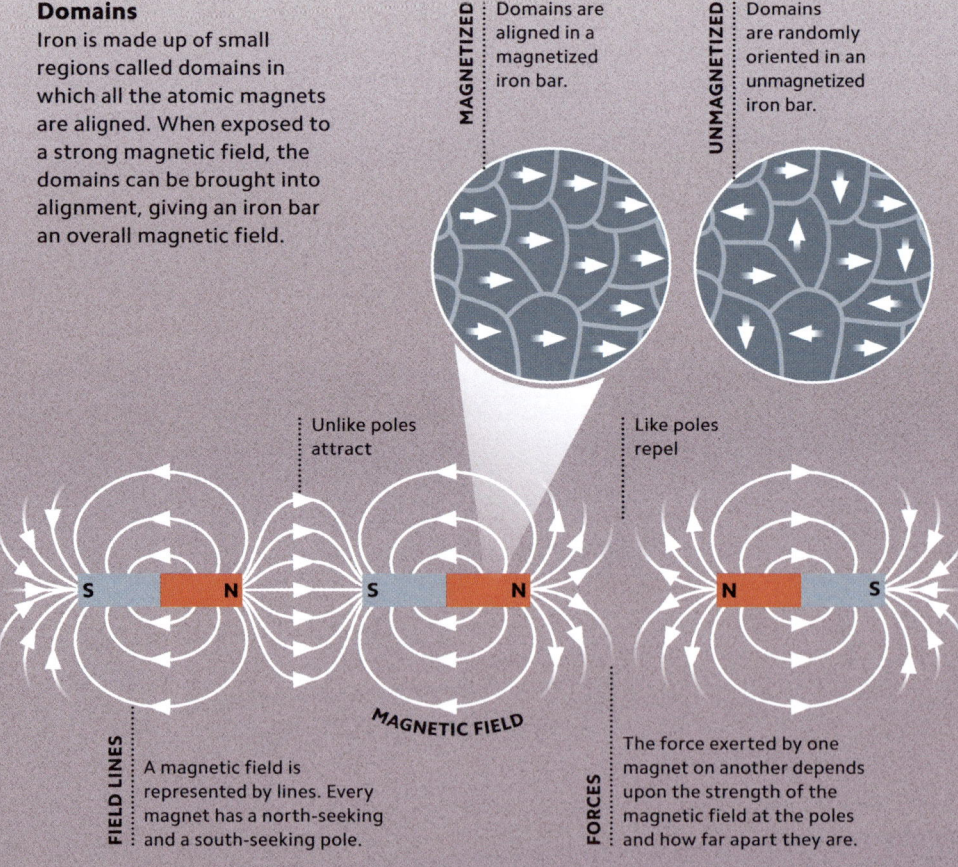

Unlike poles attract

Like poles repel

MAGNETIC FIELD

FIELD LINES A magnetic field is represented by lines. Every magnet has a north-seeking and a south-seeking pole.

FORCES The force exerted by one magnet on another depends upon the strength of the magnetic field at the poles and how far apart they are.

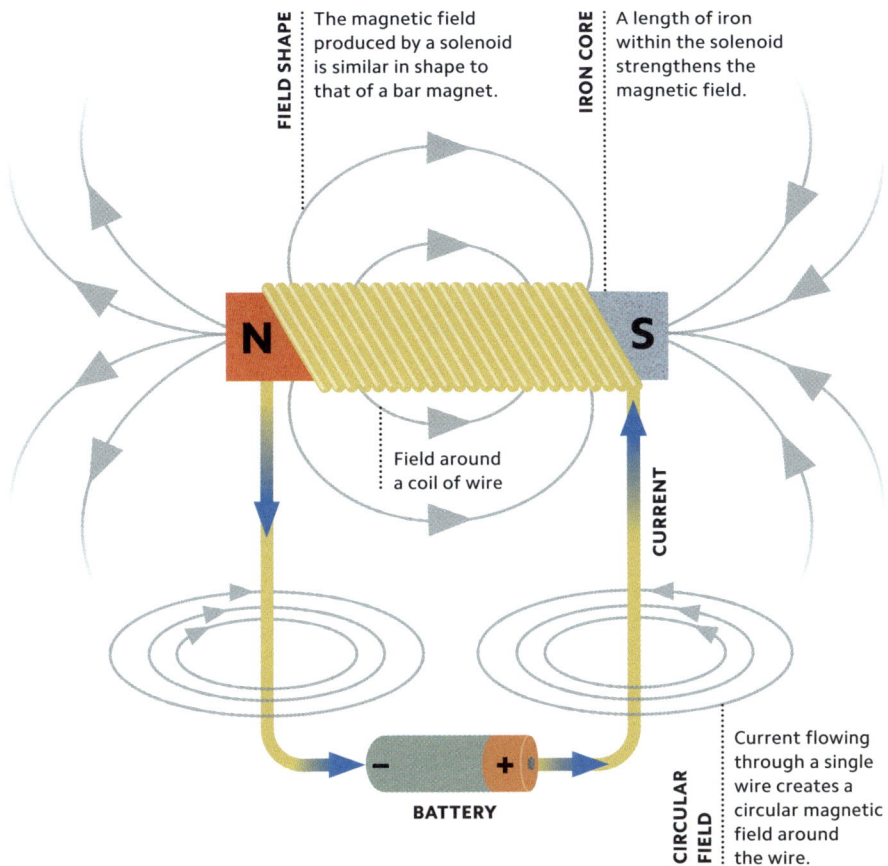

The magnetic field produced by a solenoid is similar in shape to that of a bar magnet.

A length of iron within the solenoid strengthens the magnetic field.

N

S

Field around a coil of wire

CURRENT

BATTERY

CIRCULAR FIELD
Current flowing through a single wire creates a circular magnetic field around the wire.

− +

SWITCHED ATTRACTION

Electromagnetism is the study of the relationship between electricity and magnetism. A current flowing in a wire generates a magnetic field around it. Coiling the wire to form a solenoid concentrates the field and gives the solenoid a similar field pattern to a permanent bar magnet (see opposite), but one that can be turned on and off with a switch. Introducing an iron core into the solenoid enhances the field further, producing a even stronger electromagnet.

TRANSFER OF POWER

When a conductor, such as a metal wire, is placed within a fluctuating magnetic field, a current is induced to flow in the wire. Devices such as generators and transformers (which change the voltage of an electrical supply) rely on this effect, which is called induction. A transformer consists of two coils. When an alternating current flows through the primary coil, it creates a changing magnetic field, which induces a voltage in the secondary coil. The number of turns of wire in each coil determines the voltage change.

Step up, step down
Transformers that increase voltage are known as step-up; those that decrease voltage are step-down. They are both used in the transmission network that delivers electricity to homes.

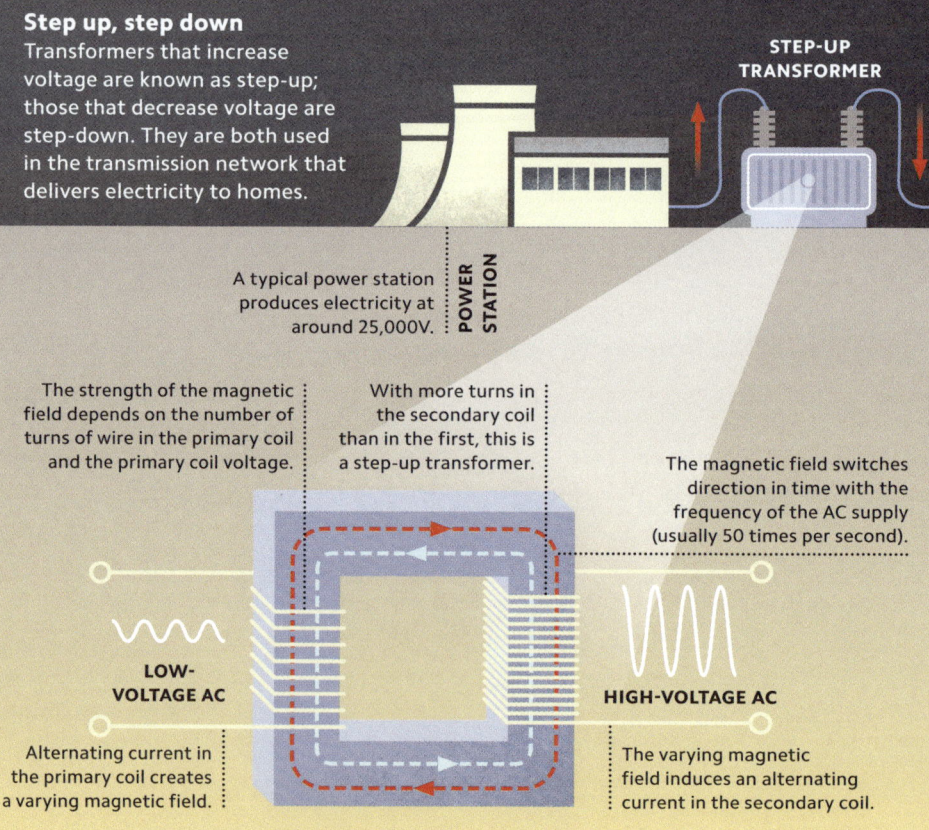

STEP-UP TRANSFORMER

A typical power station produces electricity at around 25,000V.

POWER STATION

The strength of the magnetic field depends on the number of turns of wire in the primary coil and the primary coil voltage.

With more turns in the secondary coil than in the first, this is a step-up transformer.

The magnetic field switches direction in time with the frequency of the AC supply (usually 50 times per second).

LOW-VOLTAGE AC

Alternating current in the primary coil creates a varying magnetic field.

HIGH-VOLTAGE AC

The varying magnetic field induces an alternating current in the secondary coil.

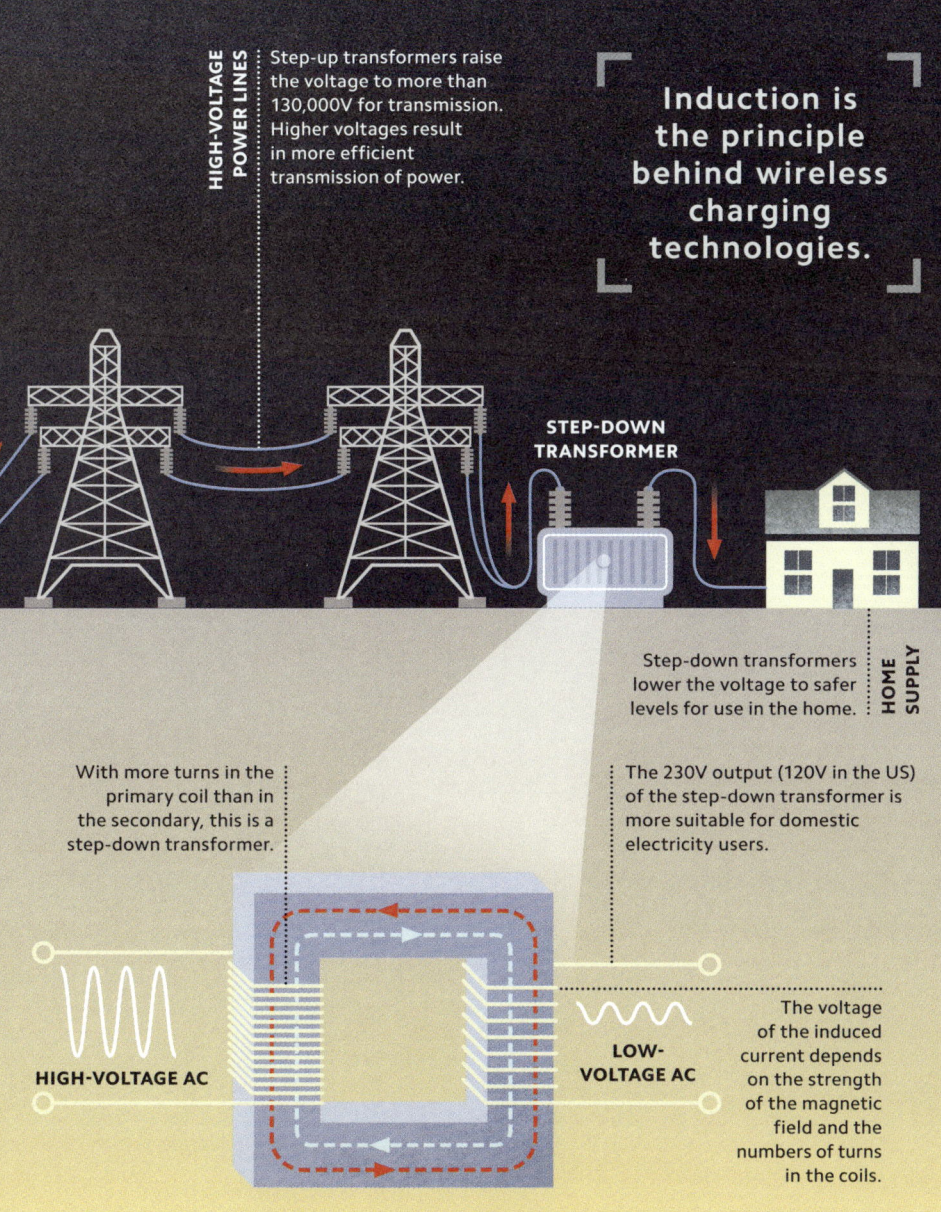

HIGH-VOLTAGE POWER LINES

Step-up transformers raise the voltage to more than 130,000V for transmission. Higher voltages result in more efficient transmission of power.

Induction is the principle behind wireless charging technologies.

STEP-DOWN TRANSFORMER

HOME SUPPLY

Step-down transformers lower the voltage to safer levels for use in the home.

With more turns in the primary coil than in the secondary, this is a step-down transformer.

The 230V output (120V in the US) of the step-down transformer is more suitable for domestic electricity users.

HIGH-VOLTAGE AC

LOW-VOLTAGE AC

The voltage of the induced current depends on the strength of the magnetic field and the numbers of turns in the coils.

MAGNETIC MOTION

Motors and generators put the principles of electromagnetism (see p.81) to work. The spinning part of an electric motor – called the rotor – is an electromagnet. It spins between the poles of a permanent magnet, called the stator. Current flowing through the wiring of the rotor produces a magnetic field; the interaction between this field and that of the stator creates a turning force. The rotor moves. A generator has the same parts as a motor, but it works in reverse, converting the motion of the rotor into an electric current.

Turning force
When a current flows through the rotor, its left arm is pulled downwards while the right arm is pulled upwards, so the rotor turns.

COMMUTATOR

FLOW OF ELECTRONS

REVERSE CURRENT
After a half-turn of the rotor, magnetic forces oppose its further rotation. The commutator reverses the current through the coil so that it can continue turning.

BATTERY

Spring-loaded carbon brushes maintain contact between the commutator and the battery.

BRUSHES

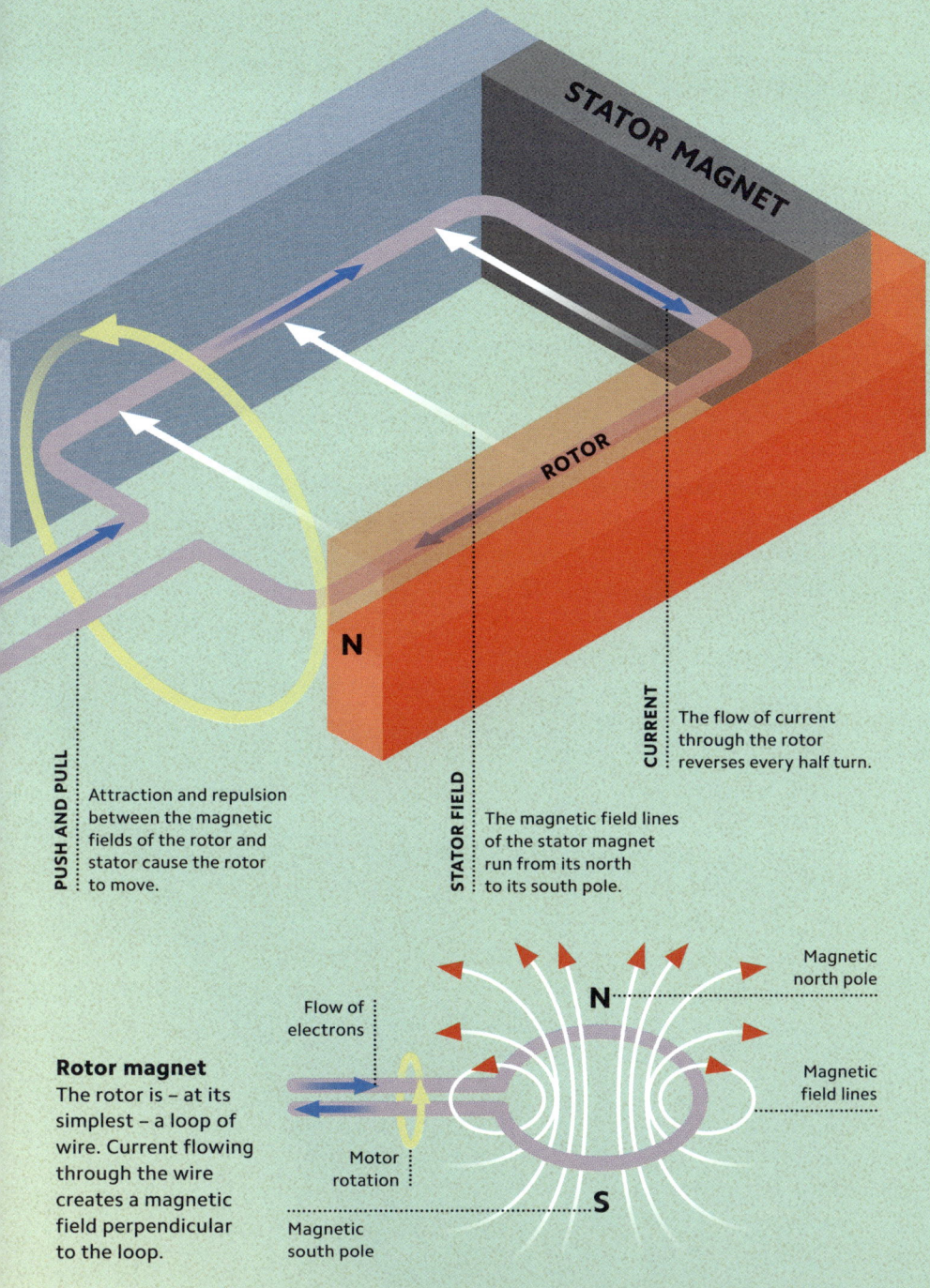

STATOR MAGNET

ROTOR

N

CURRENT
The flow of current through the rotor reverses every half turn.

PUSH AND PULL
Attraction and repulsion between the magnetic fields of the rotor and stator cause the rotor to move.

STATOR FIELD
The magnetic field lines of the stator magnet run from its north to its south pole.

Magnetic north pole

Flow of electrons

N

Magnetic field lines

Rotor magnet
The rotor is – at its simplest – a loop of wire. Current flowing through the wire creates a magnetic field perpendicular to the loop.

Motor rotation

S

Magnetic south pole

MOVING FIELDS MAKE WAVES

Like water and sound waves, electromagnetic (EM) waves are travelling waves. However, they are not disturbances of a medium (such as water or air) but of electric and magnetic fields. EM waves are produced when charged particles, such as electrons, accelerate. For example, an alternating current (see p.73) sends electrons surging back and forth in a radio antenna. This creates varying electric and magnetic fields around the antenna. As these fields expand outwards, the varying electric field induces a varying magnetic field, and vice versa. Thus a self-sustaining wave moves away from the antenna.

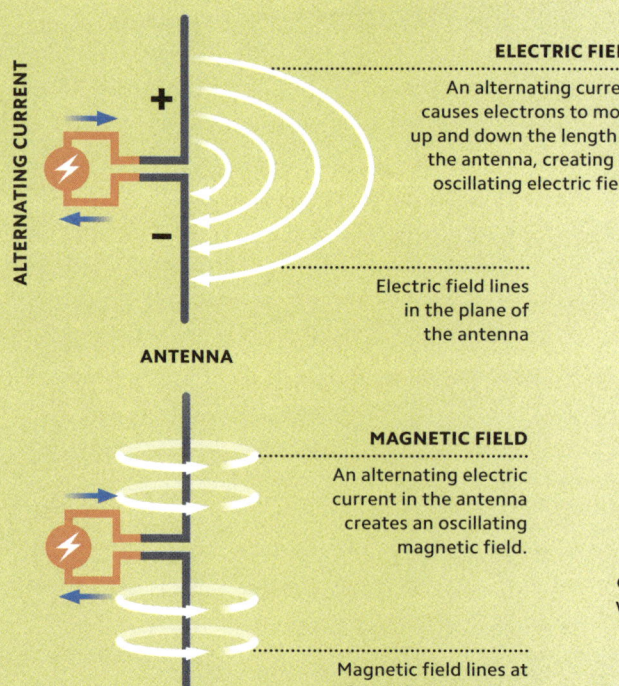

ALTERNATING CURRENT

+

−

ANTENNA

ELECTRIC FIELD
An alternating current causes electrons to move up and down the length of the antenna, creating an oscillating electric field.

Electric field lines in the plane of the antenna

MAGNETIC FIELD
An alternating electric current in the antenna creates an oscillating magnetic field.

Magnetic field lines at right angles to antenna

Creating fields
A moving charge – such as electrons flowing through a wire – will create both an electric and a magnetic field. When the current alternates, these fields expand and collapse in step with the oscillating current.

Radio transmission

A radio transmission involves making electrons oscillate in the transmitting antenna. Waves propagate outwards; some reach the receiving antenna.

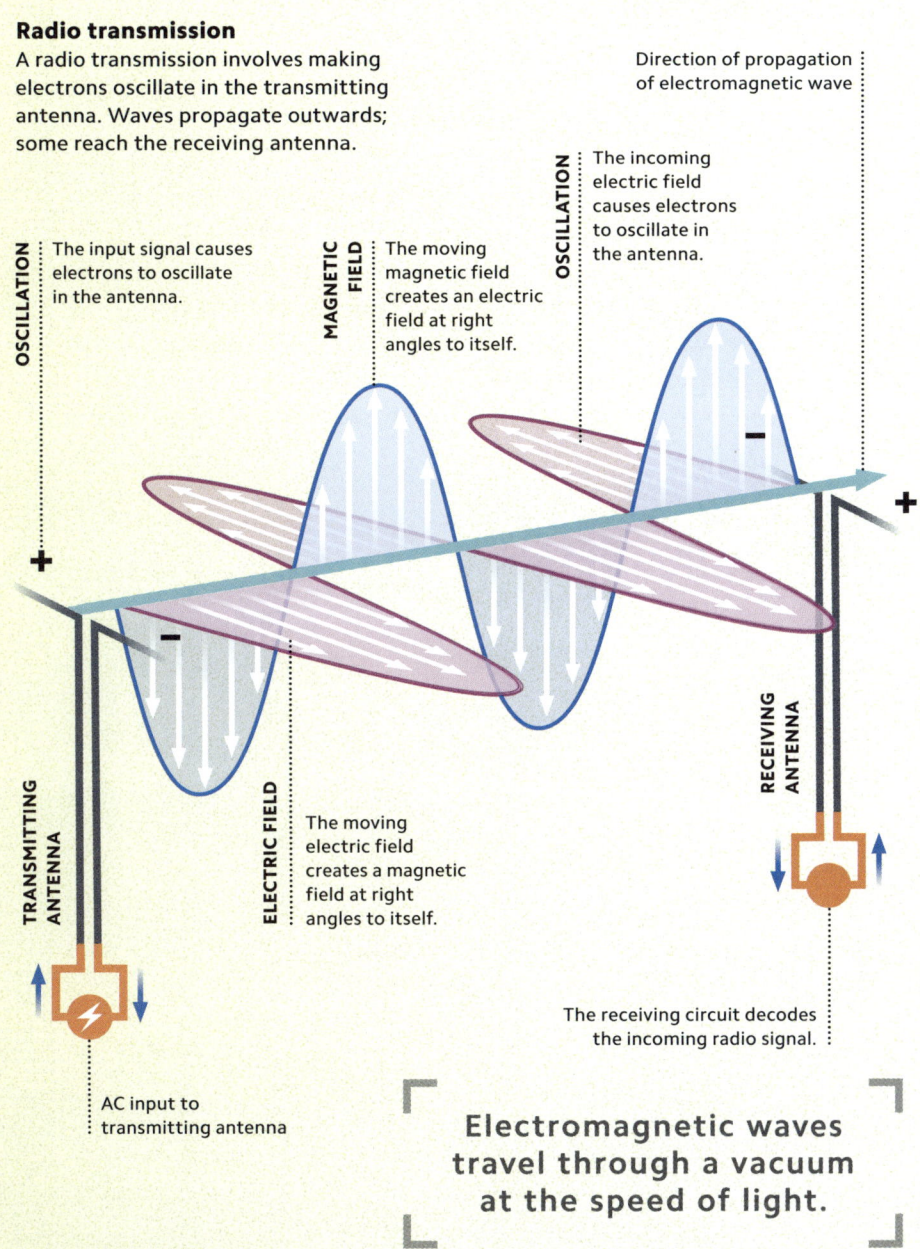

OSCILLATION

The input signal causes electrons to oscillate in the antenna.

MAGNETIC FIELD

The moving magnetic field creates an electric field at right angles to itself.

Direction of propagation of electromagnetic wave

OSCILLATION

The incoming electric field causes electrons to oscillate in the antenna.

TRANSMITTING ANTENNA

ELECTRIC FIELD

The moving electric field creates a magnetic field at right angles to itself.

RECEIVING ANTENNA

AC input to transmitting antenna

The receiving circuit decodes the incoming radio signal.

Electromagnetic waves travel through a vacuum at the speed of light.

NAVIGATING THE SPECTRUM

The space around us is crisscrossed by electromagnetic radiation, of which visible light makes up just a small portion. Different types of radiation have characteristic wavelengths that range from hundreds of kilometres to billionths of a millimetre. Electromagnetic radiation can be considered to be made up of waves or particles called photons (see pp.112–13), in which case the spectrum below represents the range of energy per photon.

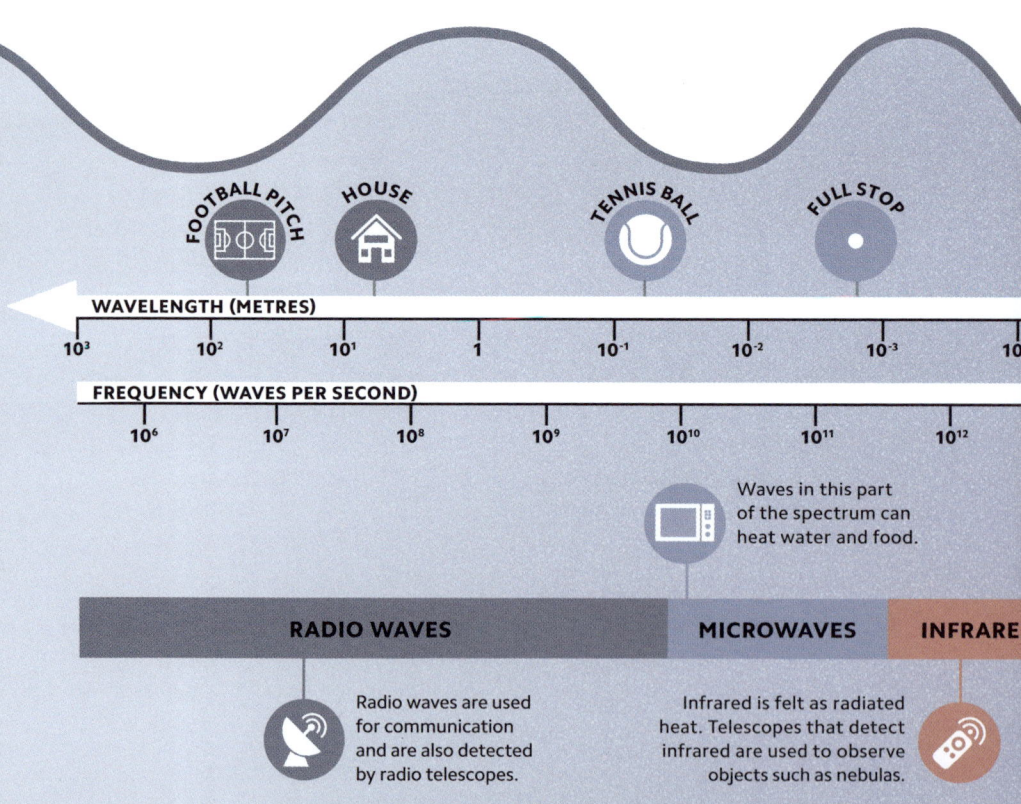

FOOTBALL PITCH HOUSE TENNIS BALL FULL STOP

WAVELENGTH (METRES)

10^3 10^2 10^1 1 10^{-1} 10^{-2} 10^{-3} 10

FREQUENCY (WAVES PER SECOND)

10^6 10^7 10^8 10^9 10^{10} 10^{11} 10^{12}

Waves in this part of the spectrum can heat water and food.

RADIO WAVES   **MICROWAVES**   **INFRARED**

Radio waves are used for communication and are also detected by radio telescopes.

Infrared is felt as radiated heat. Telescopes that detect infrared are used to observe objects such as nebulas.

Frequency and energy
The frequency of a wave increases as its wavelength shortens. The higher the frequency of a wave, the greater the energy of its photons.

> Ultraviolet light causes sunburn; higher frequencies of radiation may be deadly.

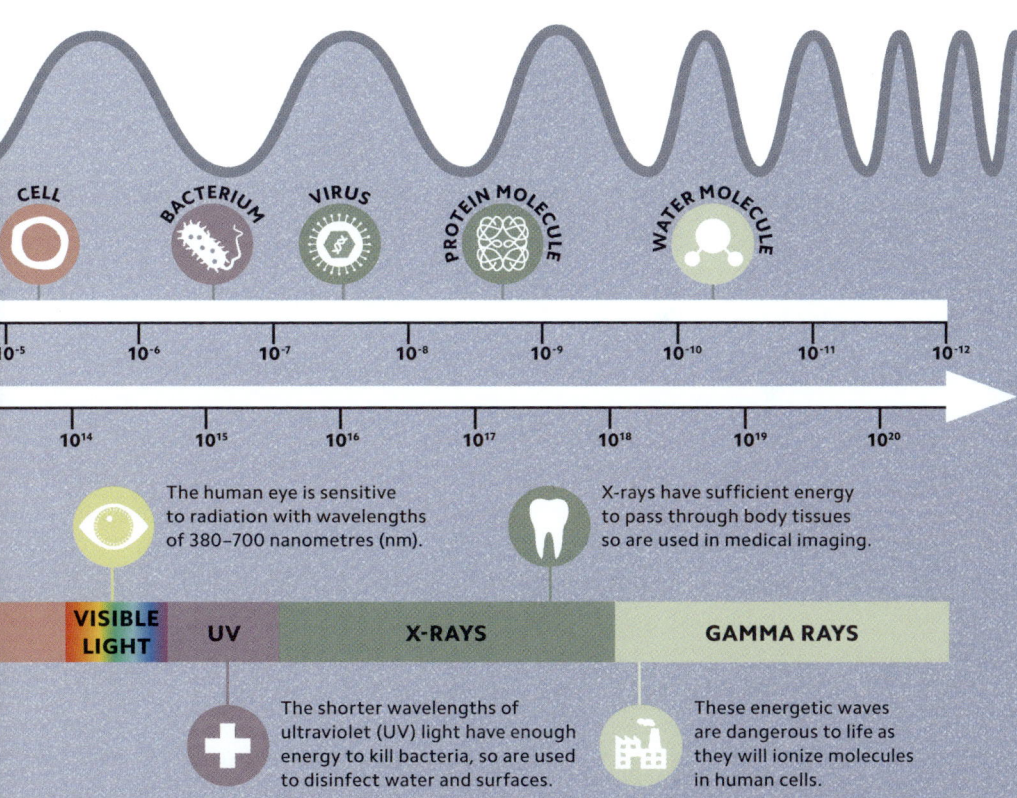

CELL

BACTERIUM

VIRUS

PROTEIN MOLECULE

WATER MOLECULE

10^{-5} 10^{-6} 10^{-7} 10^{-8} 10^{-9} 10^{-10} 10^{-11} 10^{-12}

10^{14} 10^{15} 10^{16} 10^{17} 10^{18} 10^{19} 10^{20}

The human eye is sensitive to radiation with wavelengths of 380–700 nanometres (nm).

X-rays have sufficient energy to pass through body tissues so are used in medical imaging.

VISIBLE LIGHT **UV** **X-RAYS** **GAMMA RAYS**

The shorter wavelengths of ultraviolet (UV) light have enough energy to kill bacteria, so are used to disinfect water and surfaces.

These energetic waves are dangerous to life as they will ionize molecules in human cells.

SOUND
AND LIG

H T

Although very different in nature, sound and light are both types of wave – disturbances that move away from their sources. Sound is a physical wave – a vibration that travels through a medium such as air. The frequency and amplitude of the wave determine the pitch and volume that we perceive. Light is an electromagnetic wave, which requires no medium. It is emitted by hot objects and when electrons in atoms lose specific amounts of energy. The frequency and amplitude of the wave determine the colour and brightness that we perceive. Mirrors and lenses reflect and refract light, and can form images – vital to the function of optical instruments such as telescopes and microscopes.

GOOD VIBRATIONS

Sound is how we perceive vibration. As an object vibrates, it moves to and fro; forward movement compresses the air in front, creating a ridge of raised pressure. This is followed by backward movement, which leaves a volume of decreased pressure. These waves of higher and lower pressure move away from the source of the vibration. Sound travels through the air at about 1,220kph (760mph) but much faster through a denser medium such as water or metal.

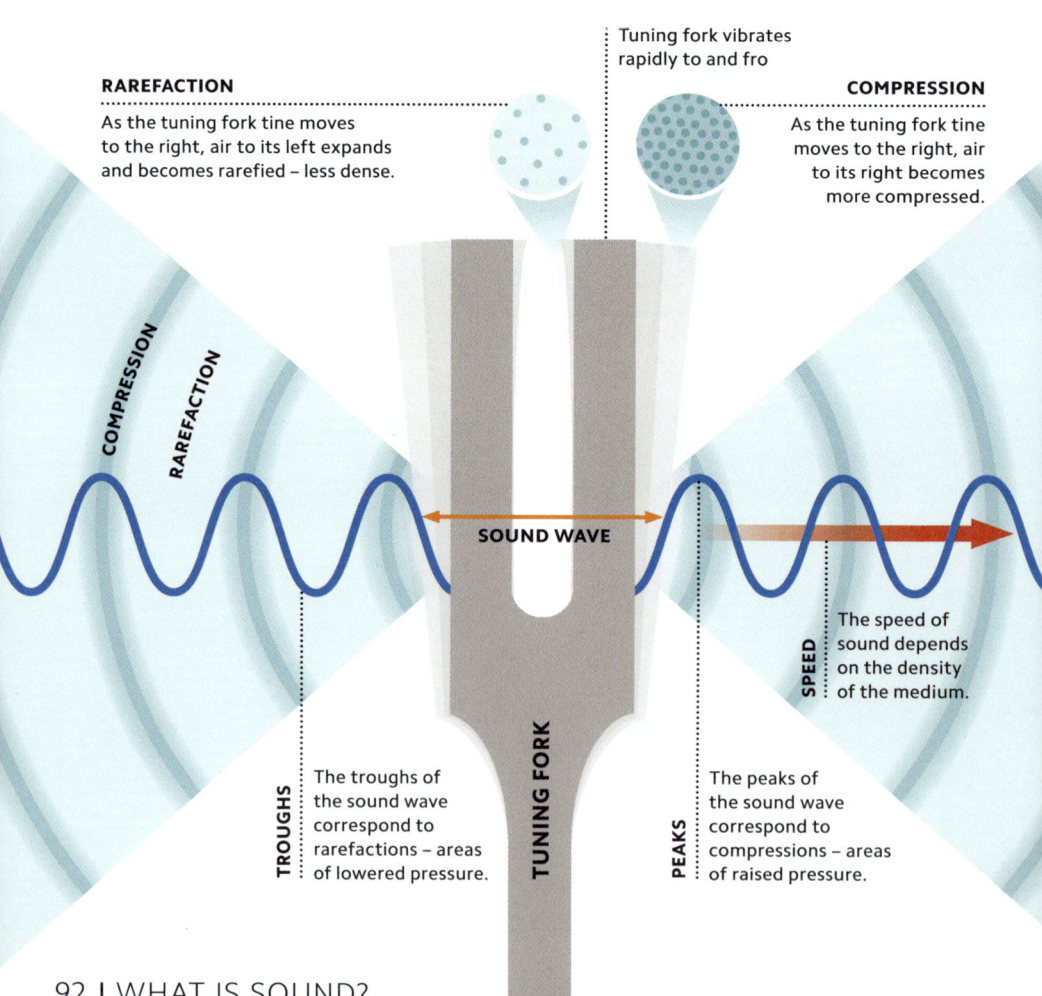

Tuning fork vibrates rapidly to and fro

RAREFACTION

As the tuning fork tine moves to the right, air to its left expands and becomes rarefied – less dense.

COMPRESSION

As the tuning fork tine moves to the right, air to its right becomes more compressed.

COMPRESSION

RAREFACTION

SOUND WAVE

SPEED

The speed of sound depends on the density of the medium.

TROUGHS

The troughs of the sound wave correspond to rarefactions – areas of lowered pressure.

TUNING FORK

PEAKS

The peaks of the sound wave correspond to compressions – areas of raised pressure.

Wavelength and amplitude

The human ear is able to hear sounds of between 20 and 20,000Hz, which correspond to wavelengths of between a few metres and a few centimetres. The wavelength and amplitude (which determines loudness) of a sound can be shown graphically, as below. In practice, a typical sound – such as that made by a musical instrument or voice – is a mixture of many wavelengths of differing amplitudes.

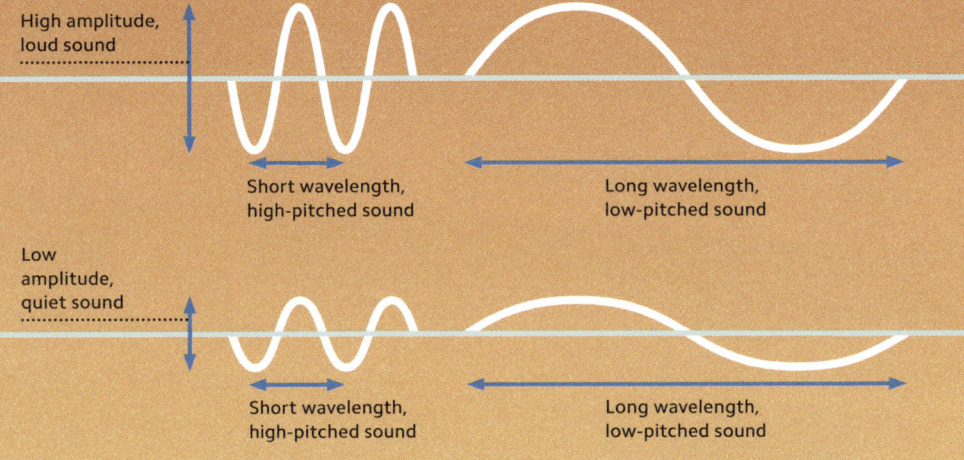

High amplitude, loud sound

Short wavelength, high-pitched sound

Long wavelength, low-pitched sound

Low amplitude, quiet sound

Short wavelength, high-pitched sound

Long wavelength, low-pitched sound

PITCH AND voLUME

The frequency of a sound is measured in hertz (Hz) and is the number of sound waves that pass a given point in one second. Higher frequencies are perceived as higher-pitched sounds. The amplitude of a sound wave is determined by how much its pressure varies above and below atmospheric pressure. Higher amplitudes carry more energy and are perceived as louder. Loudness is usually measured using the decibel (dB) scale, where sounds are given a numerical value relative to the quietest audible sound, which is defined as 0dB.

THE SOUND OF MUSIC

Plucking a guitar string makes it vibrate up and down. Since it is fixed at both ends, its vibration is constrained; only vibrations with a point of zero amplitude (node) at each end are possible. In its simplest, or fundamental, mode there is a maximum vibration (antinode) at the centre. Its frequency is determined by the string's mass, length, and tension. Other modes, called harmonics, have nodes along the length of the string, and vibrate at higher frequencies.

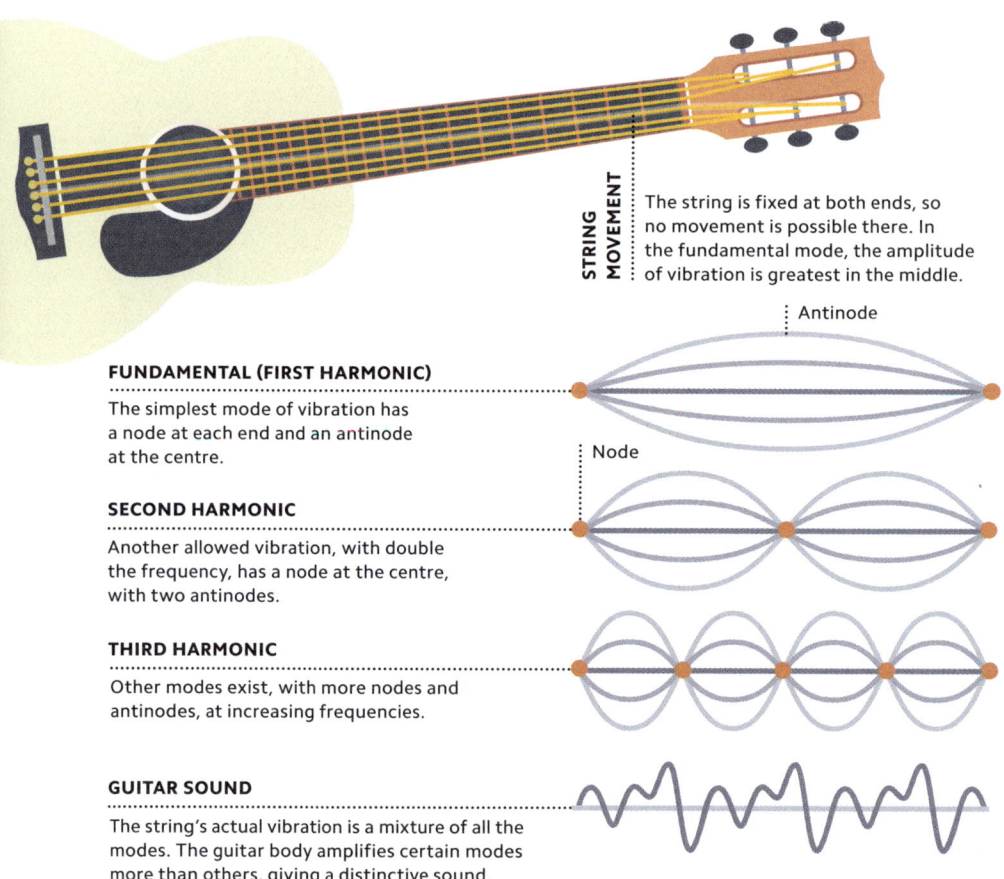

STRING MOVEMENT

The string is fixed at both ends, so no movement is possible there. In the fundamental mode, the amplitude of vibration is greatest in the middle.

Antinode

FUNDAMENTAL (FIRST HARMONIC)

The simplest mode of vibration has a node at each end and an antinode at the centre.

Node

SECOND HARMONIC

Another allowed vibration, with double the frequency, has a node at the centre, with two antinodes.

THIRD HARMONIC

Other modes exist, with more nodes and antinodes, at increasing frequencies.

GUITAR SOUND

The string's actual vibration is a mixture of all the modes. The guitar body amplifies certain modes more than others, giving a distinctive sound.

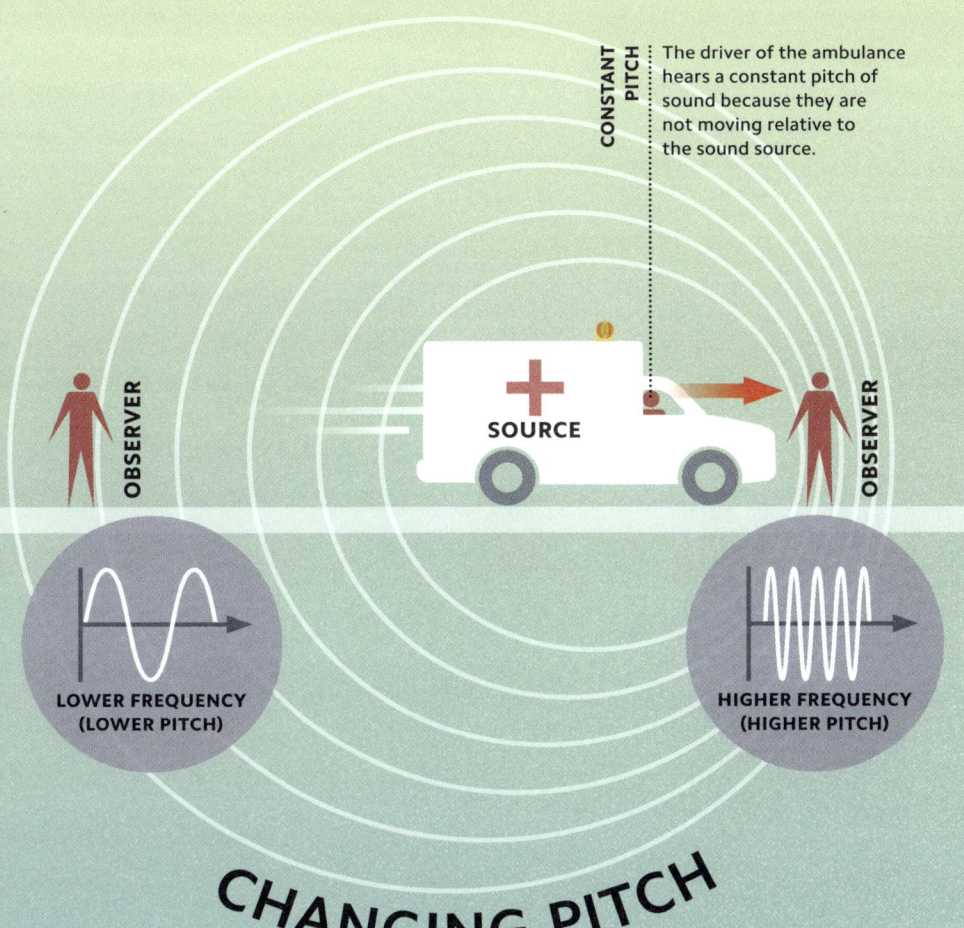

The driver of the ambulance hears a constant pitch of sound because they are not moving relative to the sound source.

OBSERVER

SOURCE

OBSERVER

LOWER FREQUENCY
(LOWER PITCH)

HIGHER FREQUENCY
(HIGHER PITCH)

CHANGING PITCH

The Doppler effect is a phenomenon common to all waves. It is most commonly encountered with sound waves. When a moving sound source (such as an ambulance with a siren) passes by an observer, the pitch (frequency) of the sound suddenly drops. This is due to relative motion: waves arrive more frequently as the ambulance is approaching, and less frequently as it is moving away. Astronomers use the Doppler effect of light to estimate the speed of faraway galaxies (see p.150).

SHINING BRIGHT

Any matter heated to a high enough temperature will glow. This is incandescence. Light can also be produced by luminescence. This occurs when electrons in atoms are excited – promoted to higher energy levels – by being heated, for example, or put into an electric field. As they fall back to a lower energy level (see pp.112–13) they emit light. A candle is a source of both incandescence and luminescence.

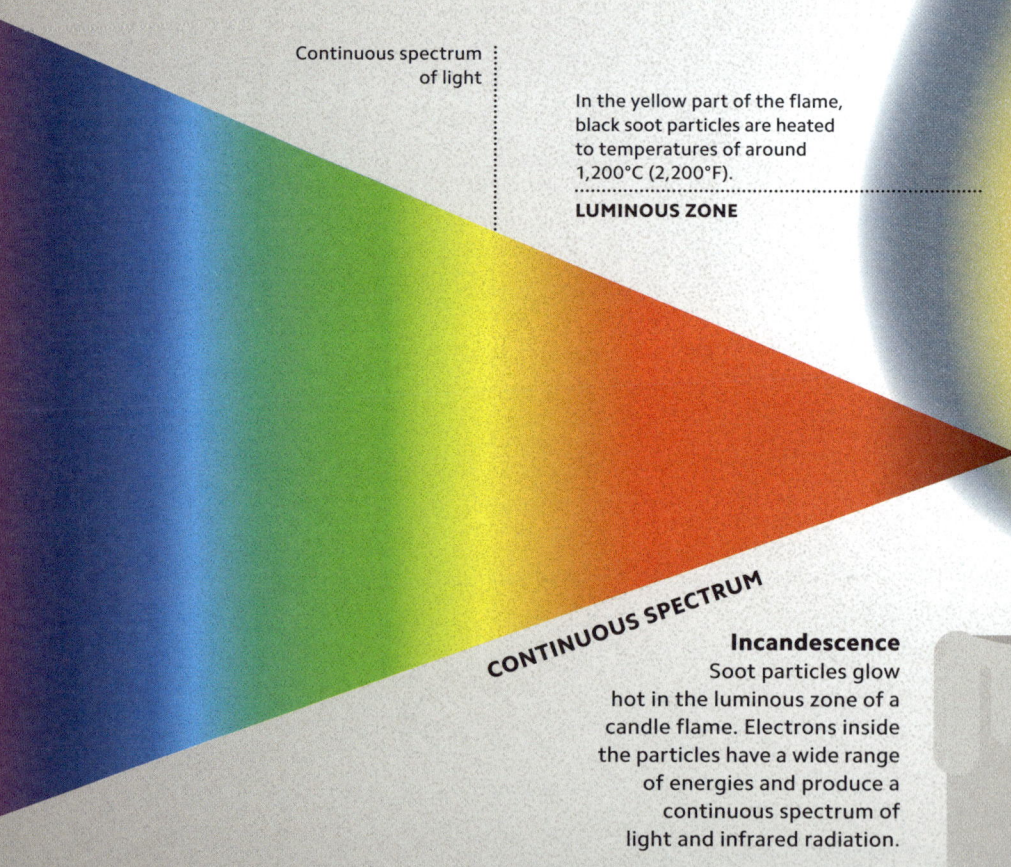

Continuous spectrum of light

In the yellow part of the flame, black soot particles are heated to temperatures of around 1,200°C (2,200°F).

LUMINOUS ZONE

CONTINUOUS SPECTRUM

Incandescence
Soot particles glow hot in the luminous zone of a candle flame. Electrons inside the particles have a wide range of energies and produce a continuous spectrum of light and infrared radiation.

Fireflies produce light through luminescence caused by a chemical reaction.

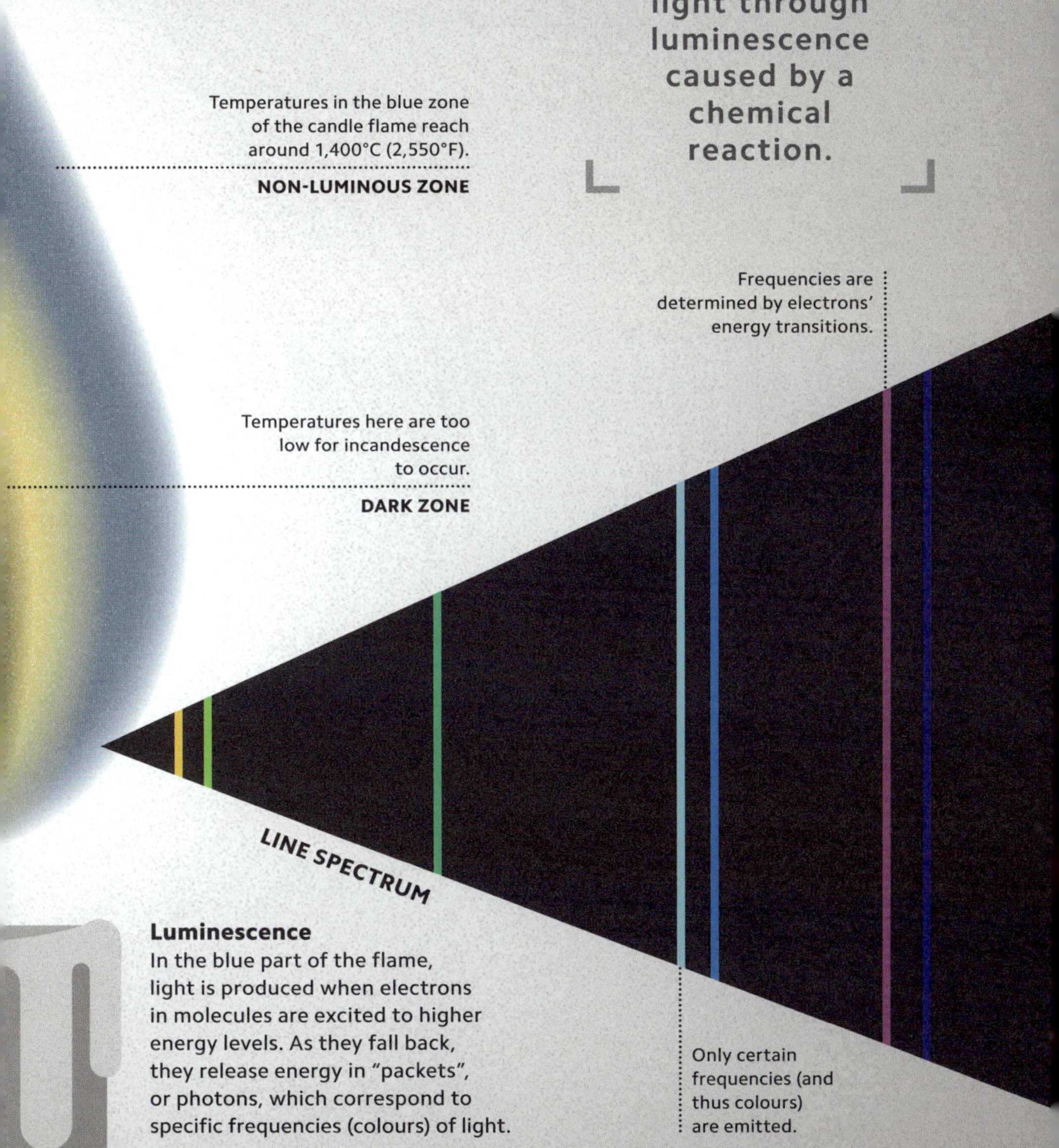
Temperatures in the blue zone of the candle flame reach around 1,400°C (2,550°F).
NON-LUMINOUS ZONE
Frequencies are determined by electrons' energy transitions.
Temperatures here are too low for incandescence to occur.
DARK ZONE
LINE SPECTRUM
Only certain frequencies (and thus colours) are emitted.

Luminescence

In the blue part of the flame, light is produced when electrons in molecules are excited to higher energy levels. As they fall back, they release energy in "packets", or photons, which correspond to specific frequencies (colours) of light.

MIRROR IMAGES

When light hits an object, some of it is absorbed, some may pass through and be refracted (see p.100), and some is reflected. There are two main kinds of reflection – specular and diffuse. Specular reflection occurs at smooth surfaces, such as mirrors, while diffuse reflection, where light scatters in many directions, occurs on objects that have irregular surfaces. How much light is reflected by an object determines how bright or dim its surface appears. A dark surface reflects almost no light, while a white one reflects nearly all of it.

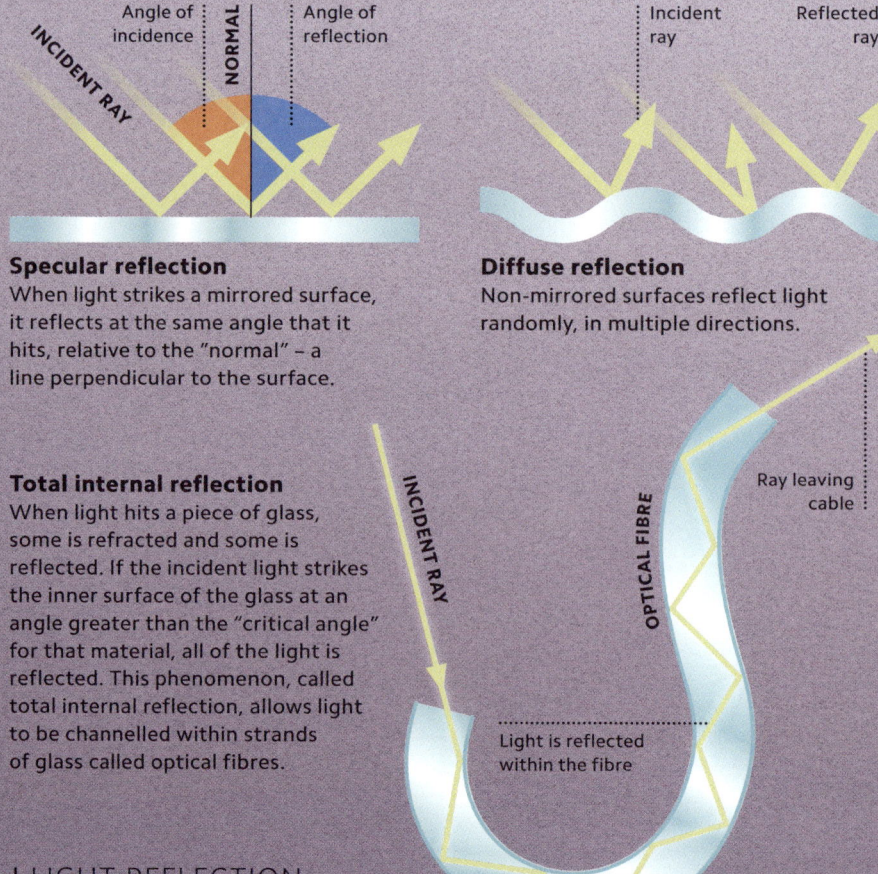

Angle of incidence | NORMAL | Angle of reflection

INCIDENT RAY

Incident ray | Reflected ray

Specular reflection
When light strikes a mirrored surface, it reflects at the same angle that it hits, relative to the "normal" – a line perpendicular to the surface.

Diffuse reflection
Non-mirrored surfaces reflect light randomly, in multiple directions.

Total internal reflection
When light hits a piece of glass, some is refracted and some is reflected. If the incident light strikes the inner surface of the glass at an angle greater than the "critical angle" for that material, all of the light is reflected. This phenomenon, called total internal reflection, allows light to be channelled within strands of glass called optical fibres.

INCIDENT RAY

OPTICAL FIBRE

Ray leaving cable

Light is reflected within the fibre

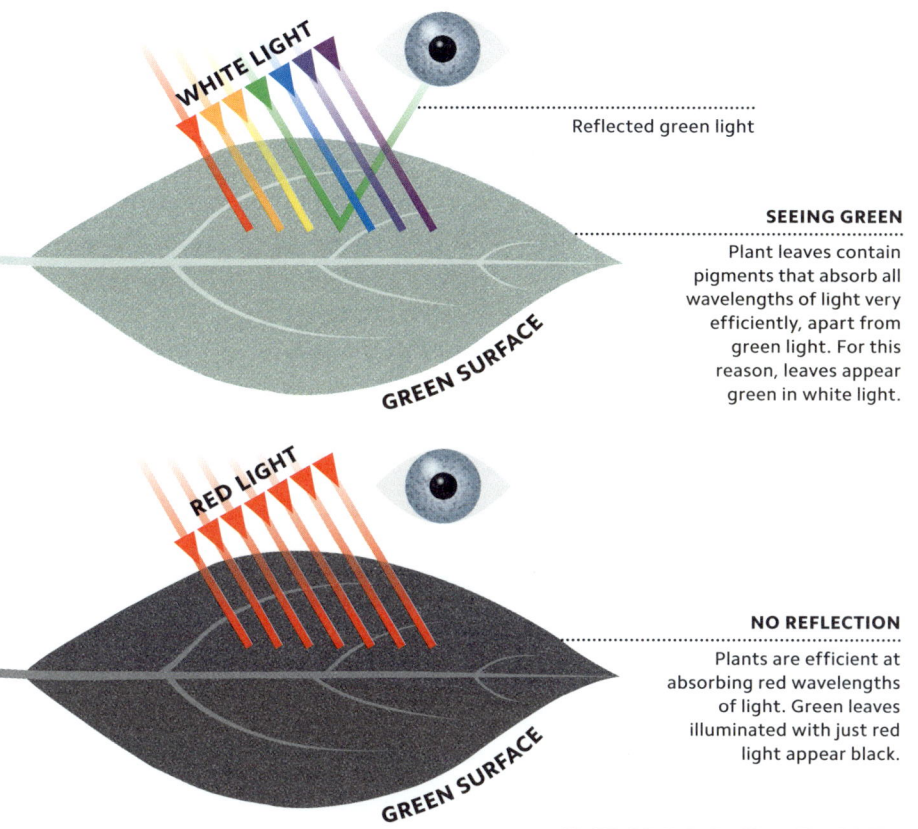

WHITE LIGHT

Reflected green light

SEEING GREEN

Plant leaves contain pigments that absorb all wavelengths of light very efficiently, apart from green light. For this reason, leaves appear green in white light.

GREEN SURFACE

RED LIGHT

NO REFLECTION

Plants are efficient at absorbing red wavelengths of light. Green leaves illuminated with just red light appear black.

GREEN SURFACE

SEEING RED

White light is a mixture of all the colours in the visible spectrum. Red light has the longest wavelength, while violet has the shortest. When light reflects from a surface, or passes through a material, light is absorbed, with some wavelengths absorbed more than others. Those not absorbed give the object its colour. Materials may also scatter light that passes through them; some wavelengths are scattered more than others. Air molecules scatter light from the blue end of the spectrum more than the red – so making the sky appear blue.

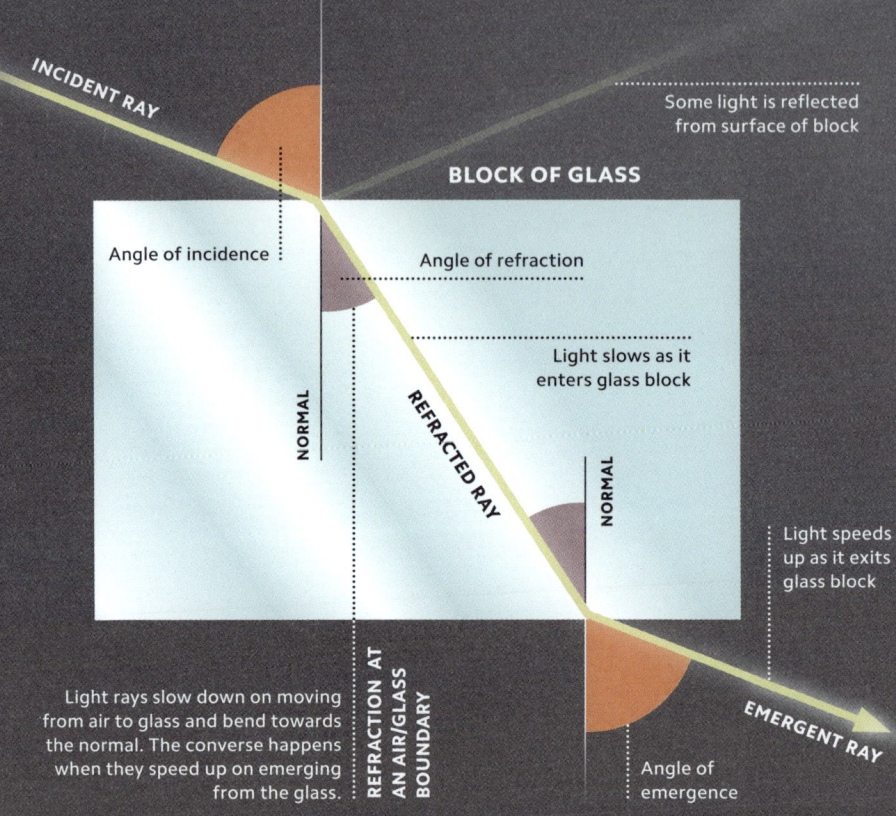

INCIDENT RAY

Some light is reflected
from surface of block

BLOCK OF GLASS

Angle of incidence

Angle of refraction

Light slows as it
enters glass block

NORMAL

REFRACTED RAY

NORMAL

Light speeds
up as it exits
glass block

Light rays slow down on moving
from air to glass and bend towards
the normal. The converse happens
when they speed up on emerging
from the glass.

REFRACTION AT
AN AIR/GLASS
BOUNDARY

EMERGENT RAY

Angle of
emergence

BENDING LIGHT

Light moves faster through a vacuum or air than it does through
a denser material such as water or glass. When light passes from
one medium to another, it changes its speed and direction. The
amount by which it is bent, or refracted, depends on the ratio of
the speed of light in the two substances. Materials such as glass
and diamond, in which light is slowed most, are said to have a
high refractive index; air and a vacuum have a low refractive index.
Some light is always reflected at a boundary; at an internal boundary,
above a certain angle of incidence, all of it is reflected (see p.98) .

Light paths in air and glass

This diagram shows the path of a beam of white light as it passes through a glass block and then a prism. It demonstrates two optical phenomena – refraction, and also dispersion, where different colours of light are refracted by different amounts.

DISPERSION

A medium such as glass slows shorter wavelengths of light more than it does longer ones. Violet light therefore bends the most, red the least. This results in dispersion – the separation of the many wavelengths of white light.

SPECTRUM

The faces of the prism are not parallel, so light is refracted twice at different angles to separate into a spectrum.

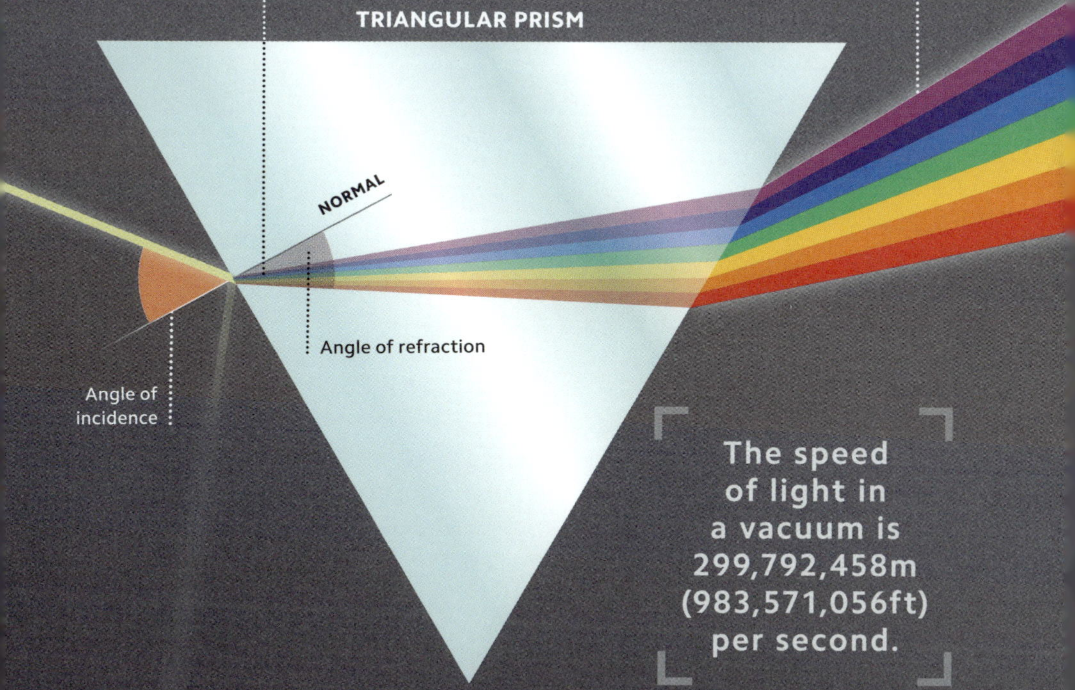

TRIANGULAR PRISM

NORMAL

Angle of refraction

Angle of incidence

The speed of light in a vacuum is 299,792,458m (983,571,056ft) per second.

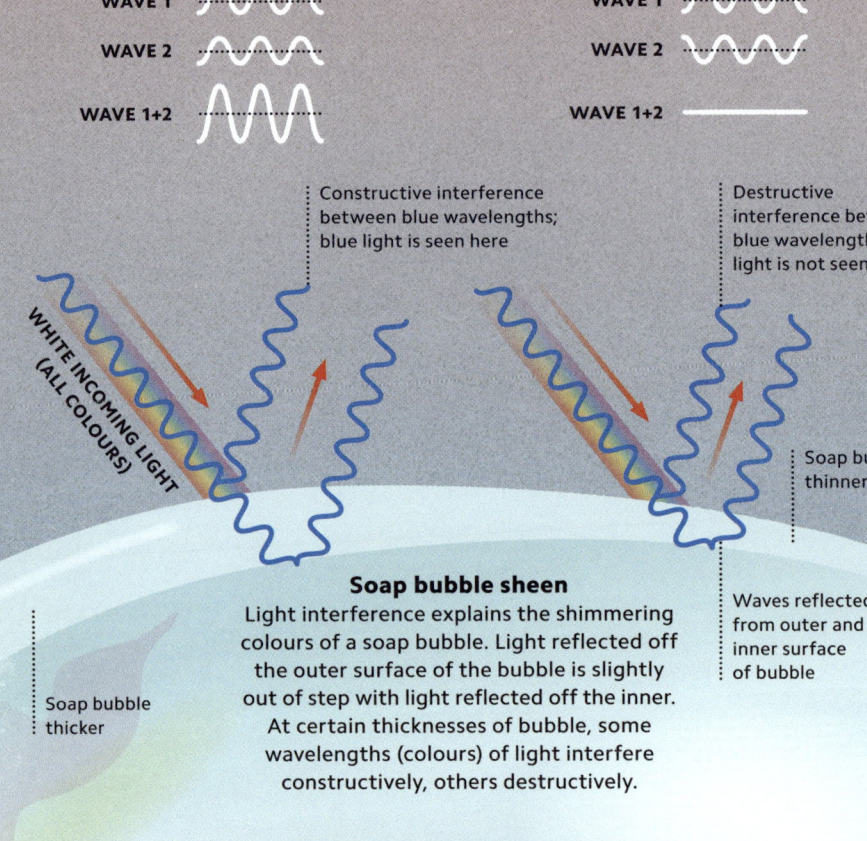

CONSTRUCTIVE

WAVE 1

WAVE 2

WAVE 1+2

DESTRUCTIVE

WAVE 1

WAVE 2

WAVE 1+2

Constructive interference between blue wavelengths; blue light is seen here

Destructive interference between blue wavelengths; blue light is not seen here

WHITE INCOMING LIGHT (ALL COLOURS)

Soap bubble thinner

Soap bubble sheen
Light interference explains the shimmering colours of a soap bubble. Light reflected off the outer surface of the bubble is slightly out of step with light reflected off the inner. At certain thicknesses of bubble, some wavelengths (colours) of light interfere constructively, others destructively.

Waves reflected from outer and inner surface of bubble

Soap bubble thicker

WAVE UPON WAVE

Interference is the phenomenon that occurs when two or more waves combine. It can be observed in all types of waves, including light, waves on water, sound waves, and other mechanical waves. Two or more waves will interfere constructively (increase their amplitude) if their peaks coincide and their troughs coincide; in this situation, the waves are said to be "in phase." If they are out of phase – with peaks of one wave coinciding with troughs of the other – they interfere destructively (cancel each other out).

SPREADING WAVES

Diffraction is the spreading out of waves when they pass an edge or emerge through a gap. The effect is particularly pronounced when a gap is about the same size as the wave's wavelength. Sound can be heard in the next room, through an open door, since its wavelengths are typically similar to the width of the door; however, light has very much shorter wavelengths, so you cannot see into the next room. Water waves diffract, too, as when waves spread out around a harbour wall.

Going round the bend
When light waves meet an obstacle, they spread out slightly. Sound spreads out more because its wavelengths are far longer.

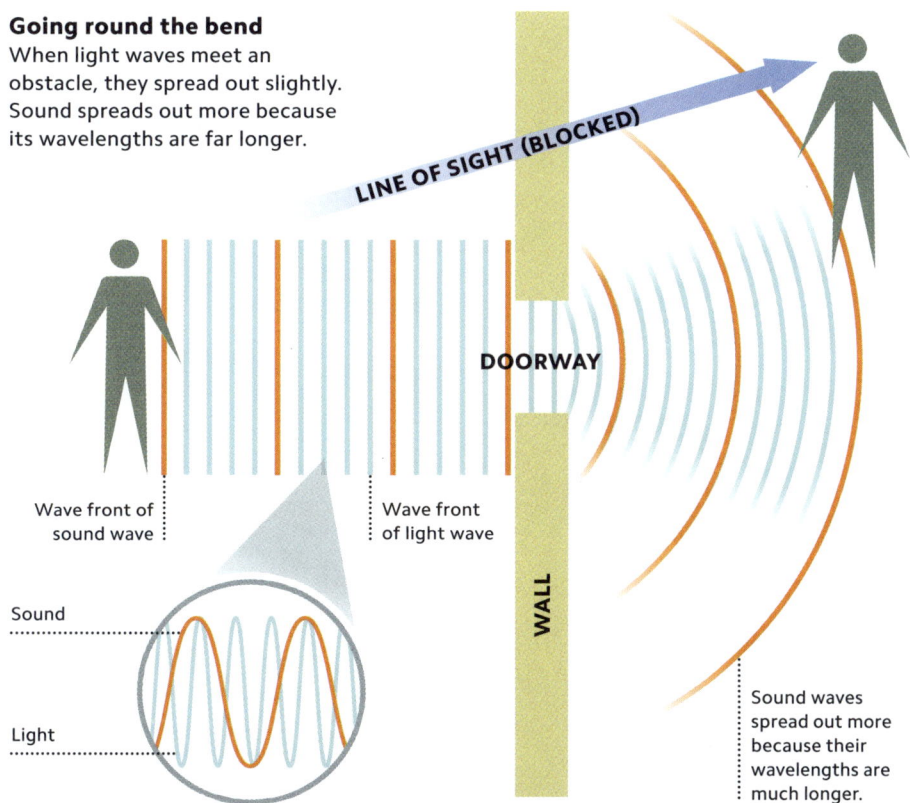

LINE OF SIGHT (BLOCKED)

DOORWAY

Wave front of sound wave

Wave front of light wave

WALL

Sound

Light

Sound waves spread out more because their wavelengths are much longer.

BENDING LIGHT

Lenses and curved mirrors can form images because they cause light rays to converge or diverge. Light rays spreading out from one point on an object can be made to converge at a single point in space by a suitable lens. This is called a real image. It can be projected onto a screen, a camera sensor, or the retina of an eye. Lenses can also form virtual images: these cannot be projected, but nevertheless make an object appear at a location in space because rays appear to come from that point. Concave lenses always form virtual images, but convex lenses can produce real and virtual images, depending on their distance from the object.

Convex lens: real image

The type of image formed by a lens depends on the distance of the object from the lens. If an object is far away from a convex lens, it will project a real image of that object. The image will be diminished in size and inverted.

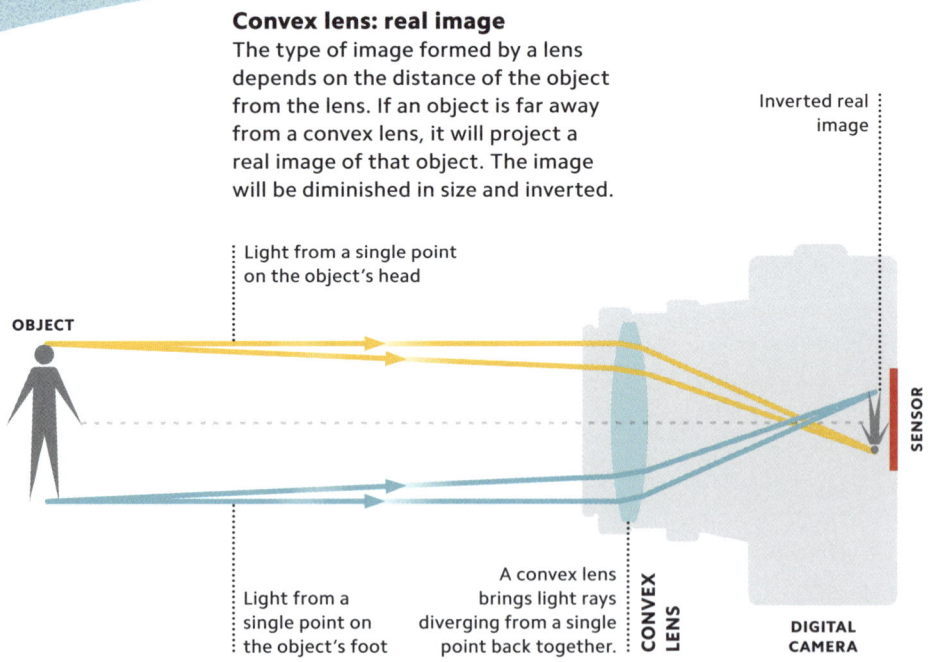

Light from a single point on the object's head

OBJECT

Inverted real image

SENSOR

Light from a single point on the object's foot

A convex lens brings light rays diverging from a single point back together.

CONVEX LENS

DIGITAL CAMERA

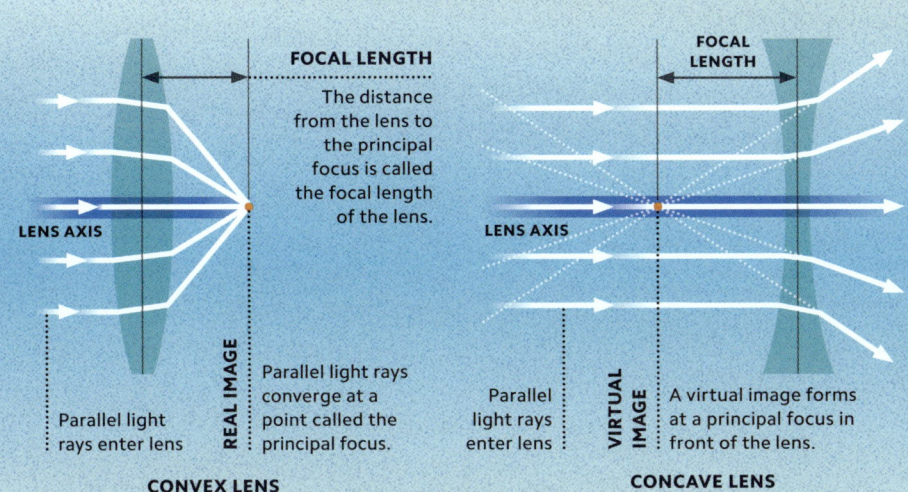

FOCAL LENGTH

The distance from the lens to the principal focus is called the focal length of the lens.

LENS AXIS

REAL IMAGE

Parallel light rays converge at a point called the principal focus.

Parallel light rays enter lens

CONVEX LENS

FOCAL LENGTH

LENS AXIS

Parallel light rays enter lens

VIRTUAL IMAGE

A virtual image forms at a principal focus in front of the lens.

CONCAVE LENS

Convex lens: virtual image

If a convex lens is closer to an object than the focal length, it behaves as a magnifying glass. The light appears to originate from a point further away from the lens than the object itself.

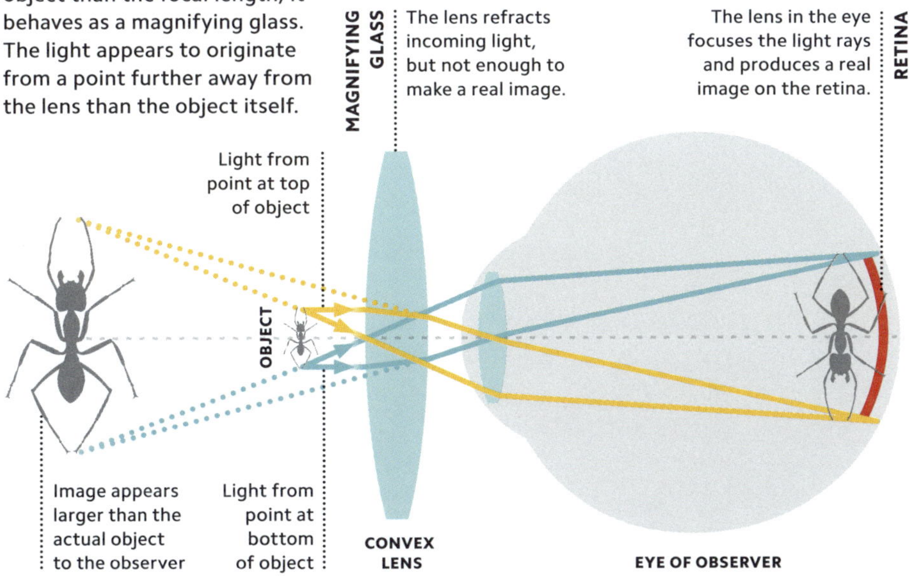

MAGNIFYING GLASS

The lens refracts incoming light, but not enough to make a real image.

RETINA

The lens in the eye focuses the light rays and produces a real image on the retina.

Light from point at top of object

OBJECT

Image appears larger than the actual object to the observer

Light from point at bottom of object

CONVEX LENS

EYE OF OBSERVER

SECOND SIGHT

Optical instruments use lenses, mirrors, or a combination of both to produce magnified images. Microscopes produce magnified images of very small objects, while telescopes produce images of distant objects, effectively bringing them closer. In both kinds of instrument, the first lens or mirror that the incoming light meets is called the objective. It produces an inverted image before an eyepiece lens alters the light path to enlarge that image.

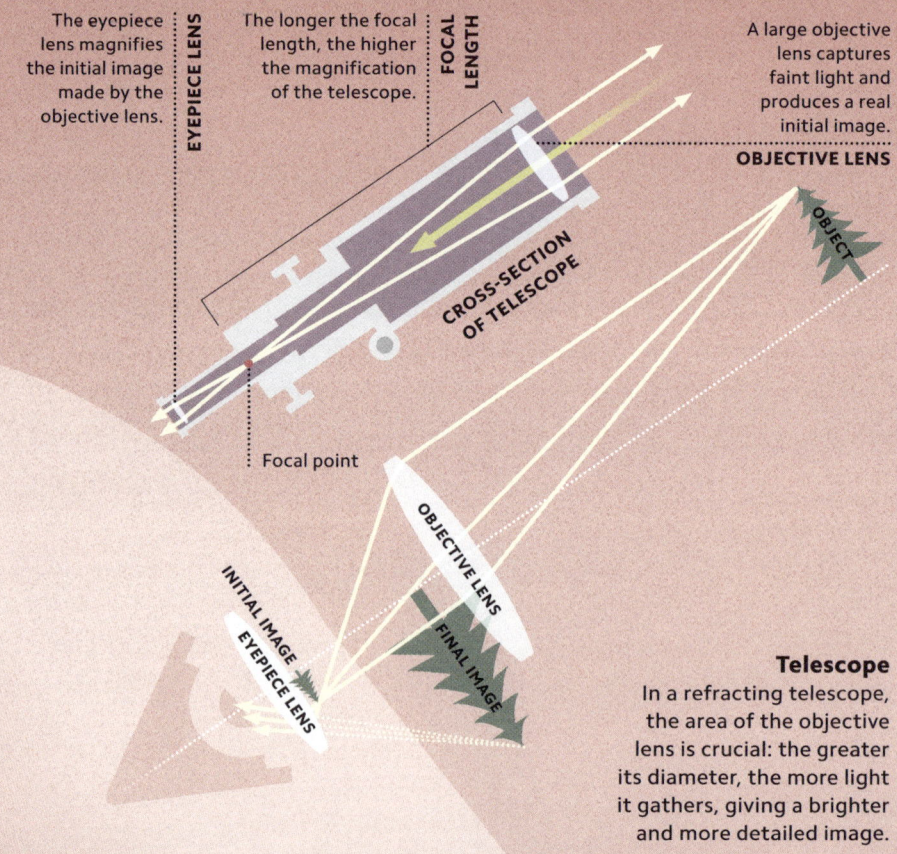

The eyepiece lens magnifies the initial image made by the objective lens.

EYEPIECE LENS

The longer the focal length, the higher the magnification of the telescope.

FOCAL LENGTH

A large objective lens captures faint light and produces a real initial image.

OBJECTIVE LENS

CROSS-SECTION OF TELESCOPE

OBJECT

Focal point

OBJECTIVE LENS

INITIAL IMAGE
EYEPIECE LENS

FINAL IMAGE

Telescope
In a refracting telescope, the area of the objective lens is crucial: the greater its diameter, the more light it gathers, giving a brighter and more detailed image.

The most powerful optical (light) microscope can magnify an object by 1,000 times.

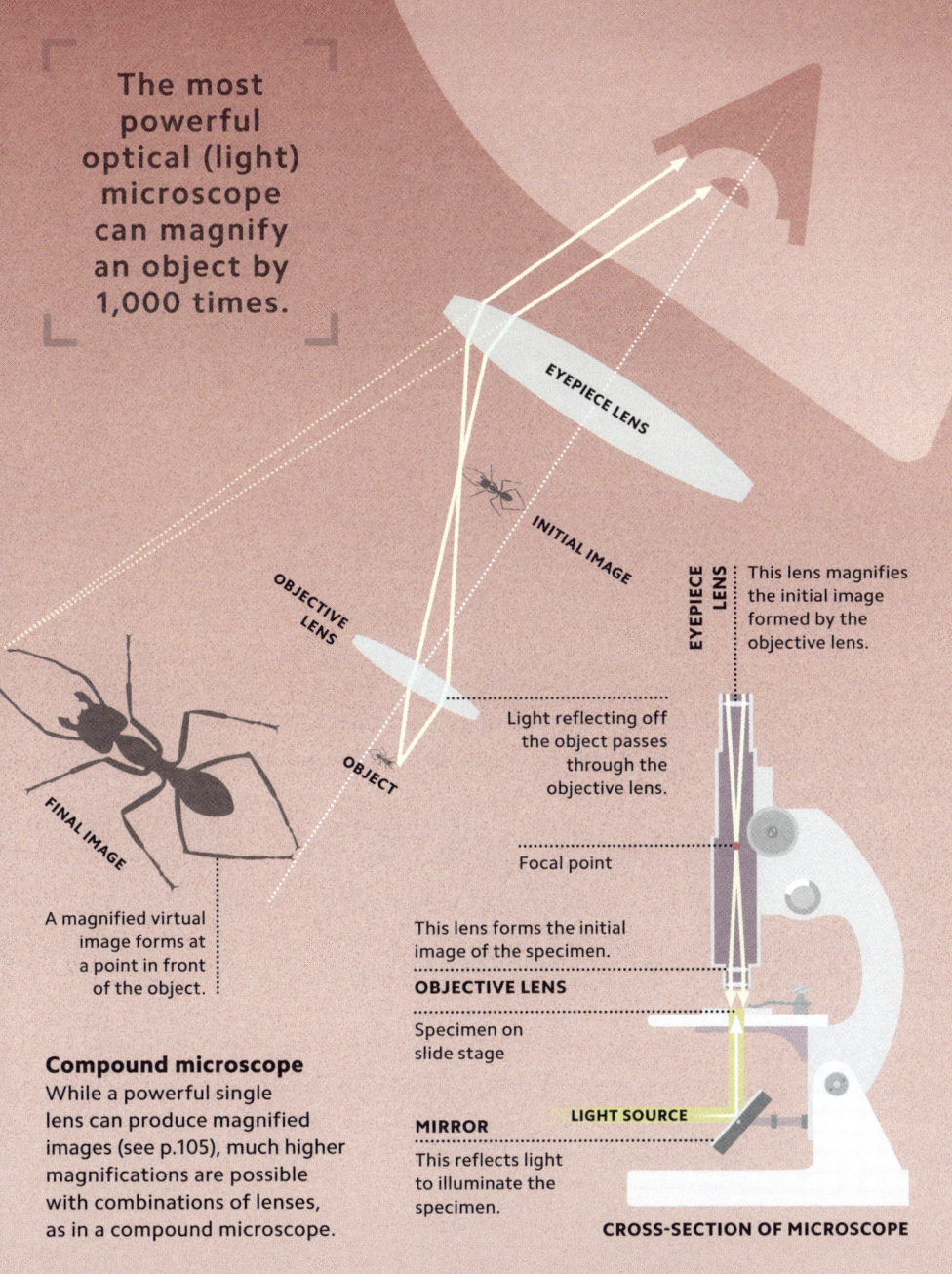

EYEPIECE LENS

INITIAL IMAGE

OBJECTIVE LENS

OBJECT

FINAL IMAGE

A magnified virtual image forms at a point in front of the object.

EYEPIECE LENS

This lens magnifies the initial image formed by the objective lens.

Light reflecting off the object passes through the objective lens.

Focal point

This lens forms the initial image of the specimen.

OBJECTIVE LENS

Specimen on slide stage

MIRROR

This reflects light to illuminate the specimen.

LIGHT SOURCE

CROSS-SECTION OF MICROSCOPE

Compound microscope

While a powerful single lens can produce magnified images (see p.105), much higher magnifications are possible with combinations of lenses, as in a compound microscope.

QUANT
PHYSI

U M
C S

Quantum physics describes the world at subatomic scales. It proposes that radiation is quantized, or broken down into discrete, indivisible packets called quanta, each with their own frequencies, wavelengths, and energies. According to quantum physics, subatomic particles behave in ways that are not like anything we are familiar with in our macroscopic world. It also suggests that uncertainty is an inherent property of the Universe. Yet it is remarkably successful at explaining the nature of light, the structure of the atom, and the interaction of light and matter. It can also be used to identify the composition of distant stars or even build devices from MRI scanners to quantum computers.

PARTICLES OF LIGHT

In 1905, Albert Einstein proposed that light does not only behave as a wave – it also behaves as a particle. This particle is called a photon. The smallest unit (or quantum) of light that can exist, a photon has no mass, no electric charge, and moves at the speed of light in a vacuum. It can be thought of as an individual packet of electromagnetic energy. The amount of energy a photon carries depends on its frequency (see pp.88-89). High-frequency photons carry more energy than those with low frequencies.

PHOTON ENERGY

In the visible part of the electromagnetic spectrum, we experience the different energies of photons as different colours of light.

PHOTON MOVEMENT

Like all particles, a photon moves in a straight line until something changes its direction.

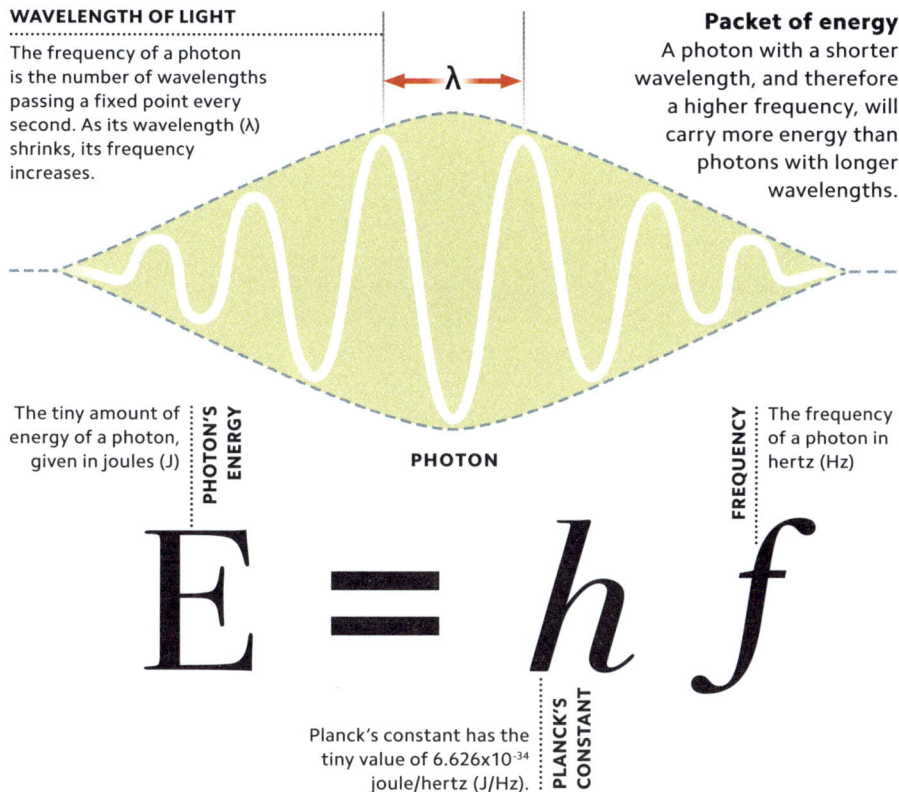

WAVELENGTH OF LIGHT

The frequency of a photon is the number of wavelengths passing a fixed point every second. As its wavelength (λ) shrinks, its frequency increases.

Packet of energy

A photon with a shorter wavelength, and therefore a higher frequency, will carry more energy than photons with longer wavelengths.

λ

The tiny amount of energy of a photon, given in joules (J)

PHOTON'S ENERGY

PHOTON

FREQUENCY

The frequency of a photon in hertz (Hz)

$$E = hf$$

Planck's constant has the tiny value of 6.626×10^{-34} joule/hertz (J/Hz).

PLANCK'S CONSTANT

FUNDAMENTAL NUMBER

The relationship between the energy of a photon and its frequency is given by the equation $E = hf$, where h is a number called Planck's constant. Any quantity of light must be a multiple of this value. This extremely small number reflects the extreme smallness of quanta of light – individual photons are so small that light appears continuous to our eyes. Planck's constant features in many of the equations of quantum mechanics. For example, the wavelength of a matter particle is Planck's constant divided by the particle's momentum.

PHOTON ENERGY

When some metals are exposed to electromagnetic radiation such as light, a process called the photoelectric effect liberates electrons from the surface of those metals. Those free electrons then form an electric current if in the presence of an electric field. However, this only happens when that light has a frequency above a certain threshold frequency; low-frequency light will not provide enough energy to liberate electrons from the metal, even if the number of photons of light directed towards it is increased. This is because the effect depends on electrons being struck by individual light quanta with sufficient energy.

LIGHT
Electromagnetic radiation falling on the metal surface can be viewed as a stream of photons – tiny packets with different amounts of energy.

COLLECTOR
When released, electrons travel through a vacuum, losing minimal kinetic energy when colliding with gas molecules. They are attracted to the collector due to its positive charge.

ELECTRON
If an electron absorbs enough energy from a photon, it escapes the metal. Any additional energy it absorbs contributes to its kinetic energy.

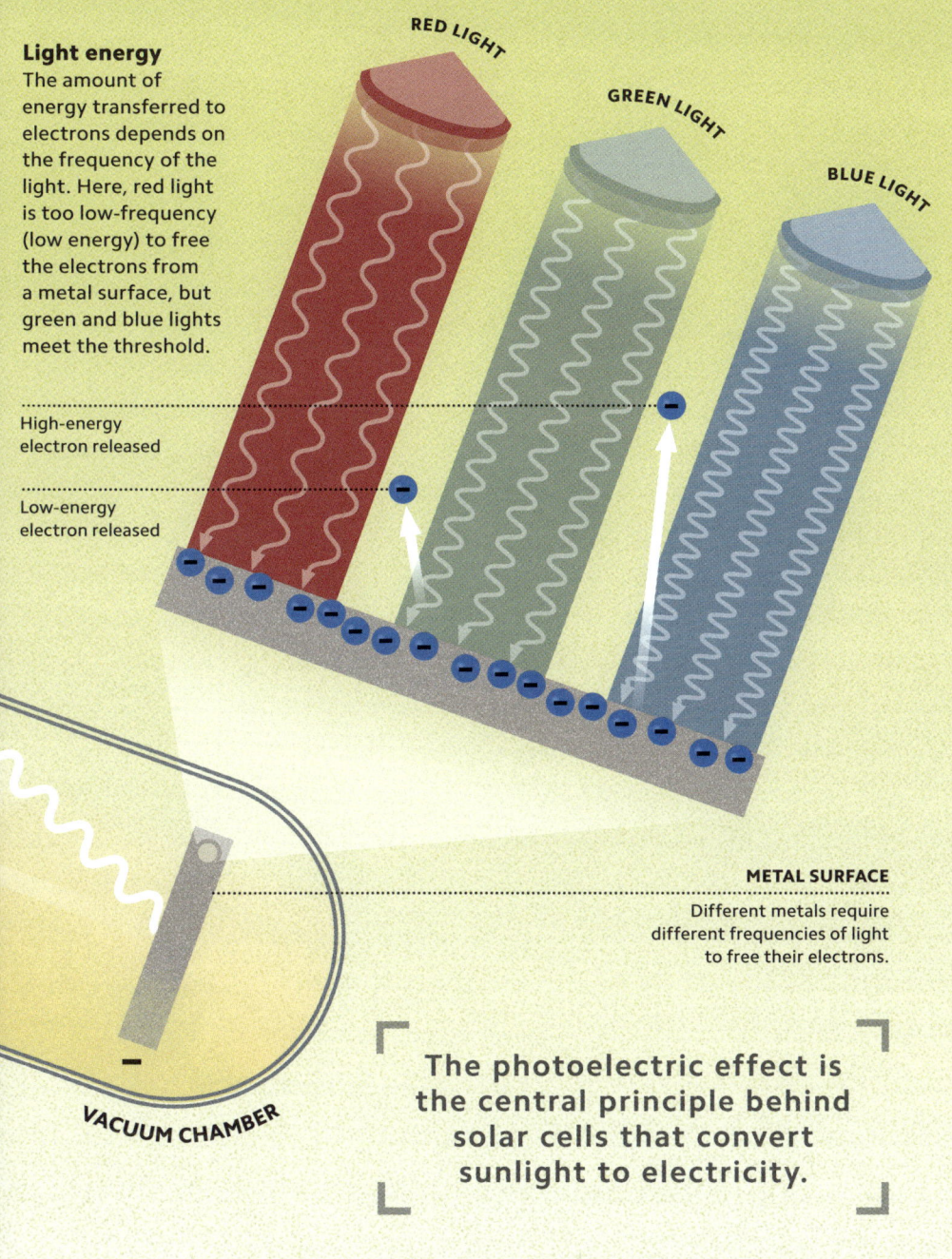

Light energy
The amount of energy transferred to electrons depends on the frequency of the light. Here, red light is too low-frequency (low energy) to free the electrons from a metal surface, but green and blue lights meet the threshold.

RED LIGHT

GREEN LIGHT

BLUE LIGHT

High-energy electron released

Low-energy electron released

METAL SURFACE
Different metals require different frequencies of light to free their electrons.

VACUUM CHAMBER

The photoelectric effect is the central principle behind solar cells that convert sunlight to electricity.

Light and matter can behave like waves
and have wavelike properties, such as
frequency and wavelength.

PARTICLE
Light and matter may also show
particlelike behaviour. For instance,
the photoelectric effect shows that
light comes in discrete packets.

TWO AT A TIME

The photoelectric effect
demonstrates that light, which
had previously been considered
strictly a wave, could also be
thought of as a particle. Later
experiments revealed that electrons
and other types of matter could
also behave like waves. Classical
concepts of waves and particles
fail to describe the behaviour of
objects in the quantum world,
where wavelike or particlelike
properties appear depending on the
circumstances of an experiment.

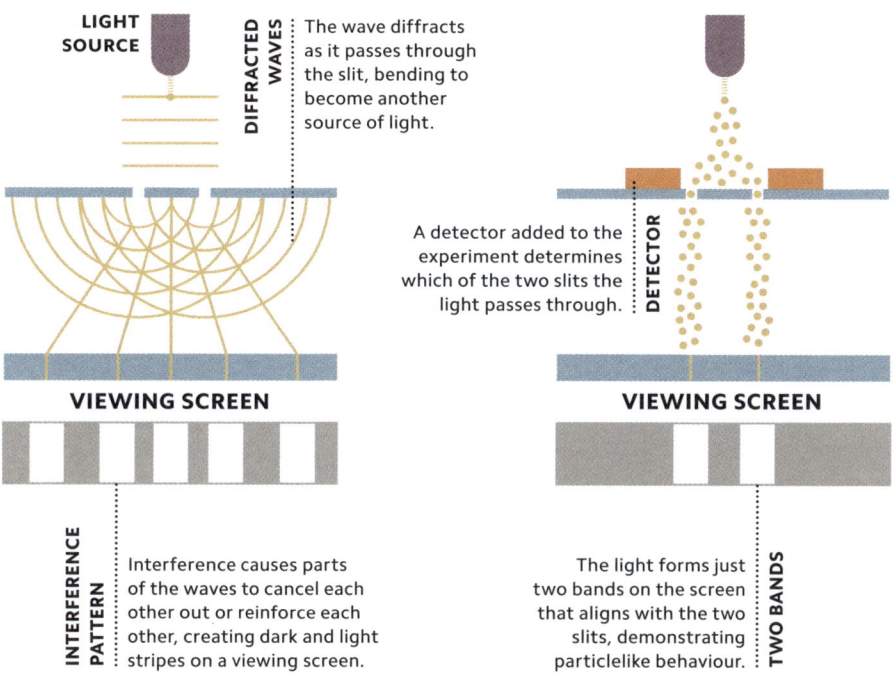

Wavelike behaviour
Light can behave as a wave – it diffracts as it passes through the slits and then interferes with itself, creating an interference pattern.

Particlelike behaviour
Light can also behave as a particle. Each particle passes through one of the two slits and strikes the screen in line with the slit it passes through.

LIGHT SOURCE

DIFFRACTED WAVES

The wave diffracts as it passes through the slit, bending to become another source of light.

A detector added to the experiment determines which of the two slits the light passes through.

DETECTOR

VIEWING SCREEN

VIEWING SCREEN

INTERFERENCE PATTERN

Interference causes parts of the waves to cancel each other out or reinforce each other, creating dark and light stripes on a viewing screen.

The light forms just two bands on the screen that aligns with the two slits, demonstrating particlelike behaviour.

TWO BANDS

THE OBSERVER EFFECT

In this famous experiment, light is directed towards a plate with two slits, illuminating a screen behind it with bands of light. The wavelike behaviour of light produces multiple bands on the screen. However, when "observed" using a detector, light displays particlelike behaviour. This suggests that measuring a quantum system has a profound effect on that system.

WHEN POSSIBILITIES COLLAPS_E

A wave function is a mathematical description of the state of a system (a system could be a particle). The value of a particle's wave function at a certain point in space and time is related to the probability of the particle being there at the time. Wave functions can also be used to describe other physical variables, such as spin.

BEFORE MEASUREMENT

The height of this wave function represents the probability of a particle occupying different positions in a two-dimensional space.

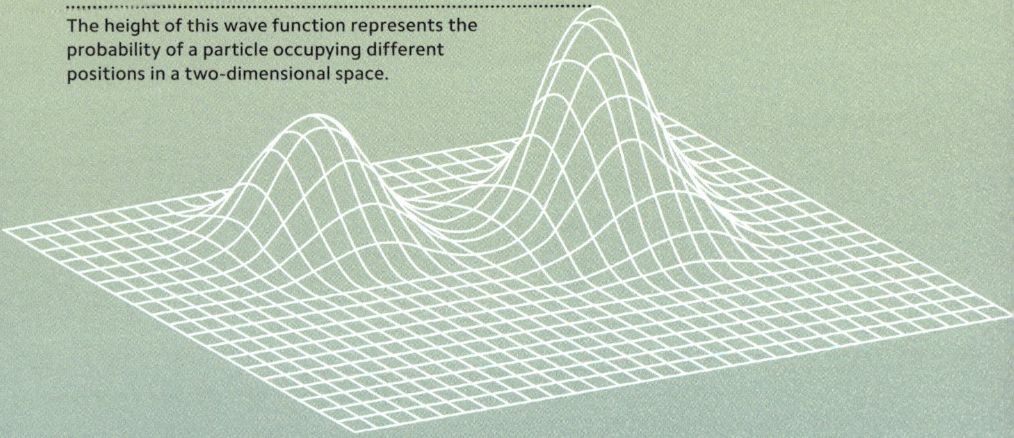

AFTER MEASUREMENT

When a measurement is made, the particle acquires definite properties, causing the wave function to "collapse" to a single state or position.

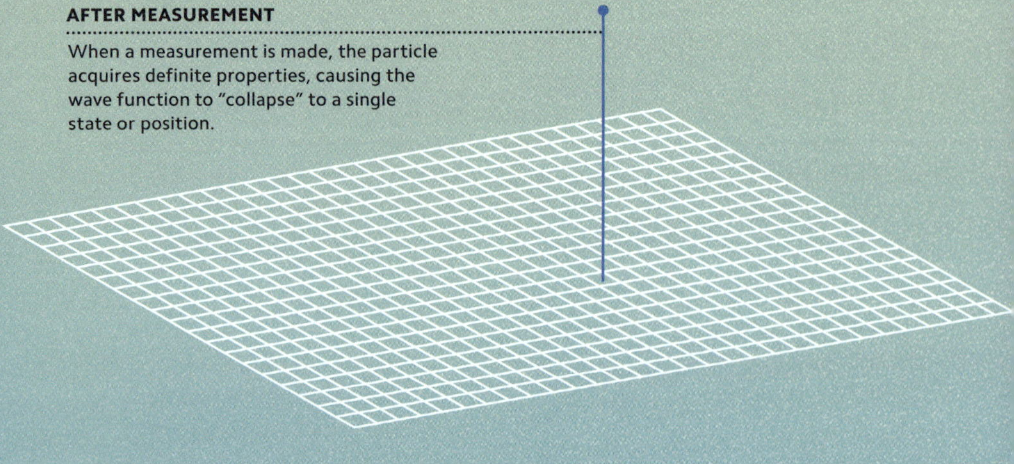

HYDROGEN

ELECTRON

SPECTRAL LINES

Each chemical element has a distinct pair of spectra, with spectral lines always occurring at the same positions. This deep red line on hydrogen's emission spectrum, known as hydrogen-alpha, corresponds to a wavelength of 656 nanometres (nm).

HYDROGEN LINES

The other lines on hydrogen's spectra are found at 486nm, 434nm, and 410nm.

Hydrogen absorption spectrum
Hydrogen has a distinctive absorption spectrum, with "missing" wavelengths in the violet, blue-green, and red parts of the spectrum.

Hydrogen emission spectrum
Hydrogen's emission spectrum is the inverse of its absorption spectrum. When an electric current is passed through hydrogen gas, it emits a blue light comprising these four wavelengths.

GIVE AND TAKE

Incandescent light will produce a continuous range of frequencies called a spectrum (see p.96). An absorption spectrum is another kind of spectrum that contains the frequencies absorbed by a material when it is exposed to radiation. This spectrum has gaps, caused by electrons jumping to a higher energy level after absorbing photons with certain wavelengths. By contrast, an emission spectrum contains only certain wavelengths, representing photons released as electrons fall to a lower energy level.

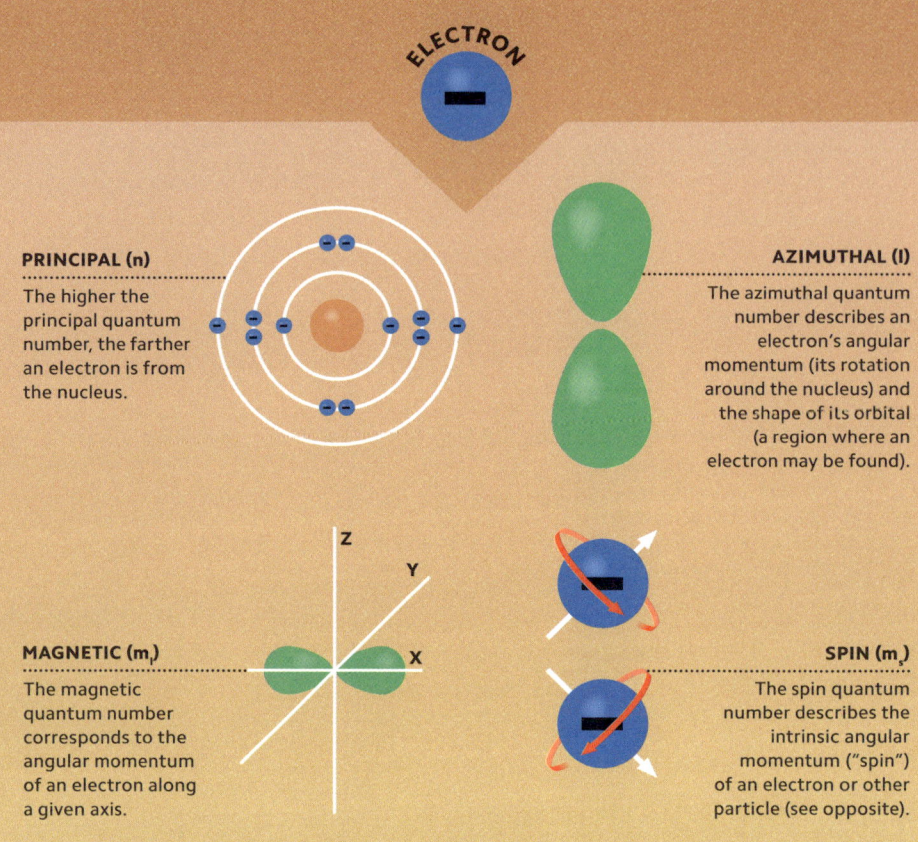

ELECTRON

PRINCIPAL (n)

The higher the principal quantum number, the farther an electron is from the nucleus.

AZIMUTHAL (l)

The azimuthal quantum number describes an electron's angular momentum (its rotation around the nucleus) and the shape of its orbital (a region where an electron may be found).

MAGNETIC (m_l)

The magnetic quantum number corresponds to the angular momentum of an electron along a given axis.

SPIN (m_s)

The spin quantum number describes the intrinsic angular momentum ("spin") of an electron or other particle (see opposite).

DEFINING STATES

Quantum numbers are quantities that describe systems that obey the laws of quantum mechanics. Different quantum numbers are needed to fully describe different systems. For example, four quantum numbers are required to fully describe an electron in a hydrogen atom: the principal, azimuthal, magnetic, and spin numbers. Other quantum numbers are needed to describe the properties of subatomic particles, such as the "flavour" of quarks (see p.34).

IN A SPIN

Spin is an intrinsic property of particles, commonly referred to as intrinsic angular momentum. This is distinct from the familiar concept of orbital angular momentum, which describes a particle's movement around a nucleus. Spin is quantized (it can only take certain discrete values), and this value is shared across all particles of the same type. For example, all photons have spin $s = 1$. Spin also interacts with a magnetic field to produce different energy levels.

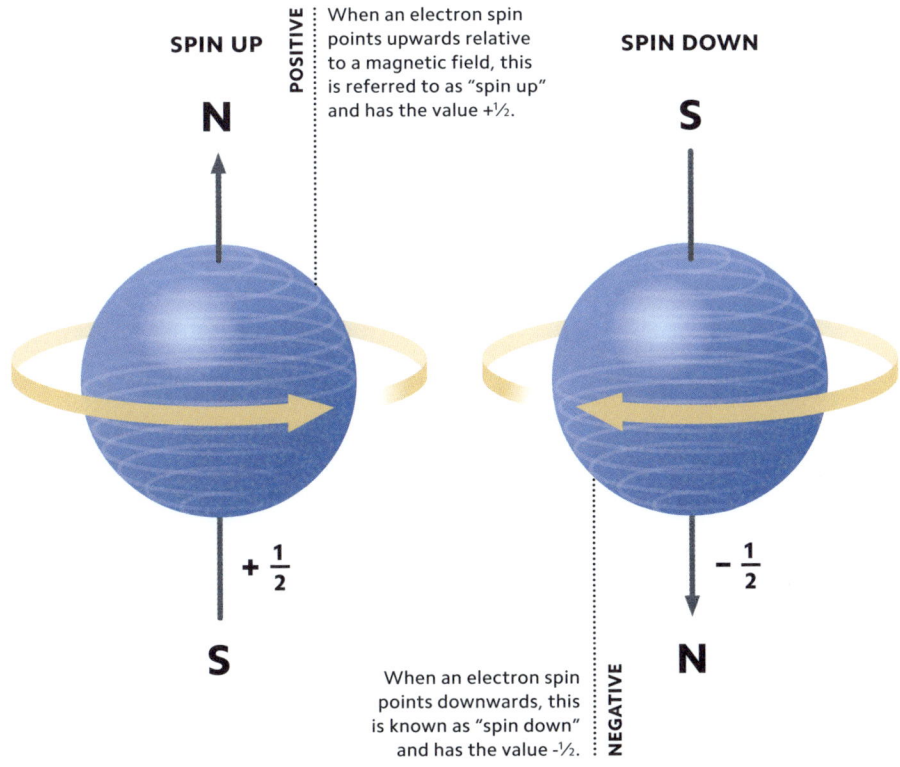

SPIN UP

POSITIVE

When an electron spin points upwards relative to a magnetic field, this is referred to as "spin up" and has the value $+\frac{1}{2}$.

SPIN DOWN

N

S

$+\frac{1}{2}$

$-\frac{1}{2}$

S

N

When an electron spin points downwards, this is known as "spin down" and has the value $-\frac{1}{2}$.

NEGATIVE

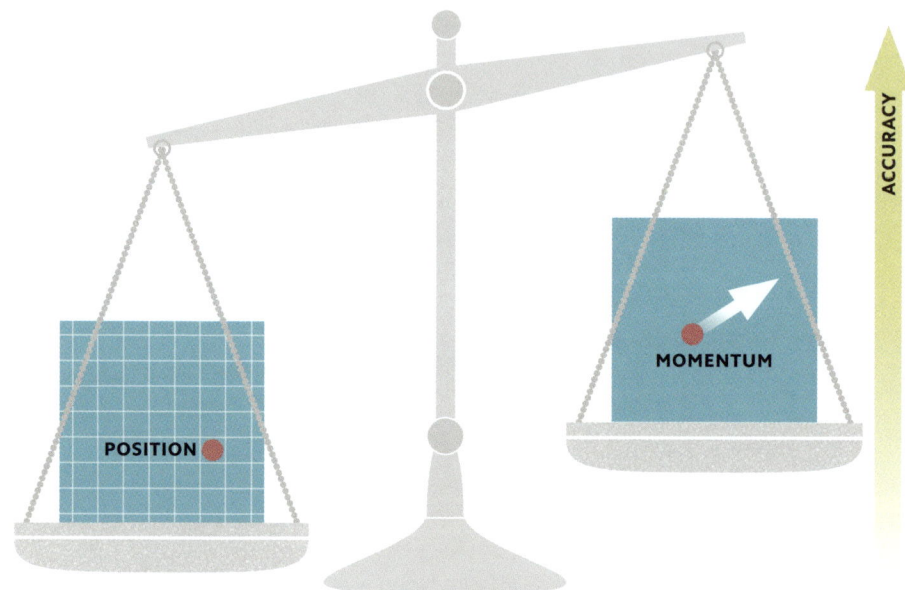

Variable pairs
The best known pair of properties affected by the uncertainty principle is position and momentum. The more accurately a particle's position is known, the less accurately its momentum can be known and vice versa. These paired properties are called conjugate variables.

ACCURACY

MOMENTUM

POSITION

KNOWING AND NOT KNOWING

In the quantum world of particles, there is a limit to how accurately certain pairs of properties – such as position and momentum – can be determined at the same time. The more precisely one of these properties is known, the less is known about the other. This is due to something called the uncertainty principle, which was discovered by German physicist Werner Heisenberg in 1927. It is made possible by wave-particle duality (see p.114) and affects all quantum objects.

FLEETING APPEARANCES

Thanks to the uncertainty principle, it is possible for virtual particles to emerge from a vacuum for extremely short periods. Virtual particles are more like disturbances in a field than ordinary particles, and they help to describe interactions between these ordinary particles. For example, the electromagnetic force between two charged particles registers as a disturbance in the electromagnetic field and can be understood as the exchange of a virtual photon.

Virtual particle
(electron)

Virtual antiparticle
(positron)

PARTICLES AND ANTIPARTICLES EMERGE

Energy and time are conjugate variables, allowing energy to emerge from "nothing" if it disappears again quickly. The more energy, the shorter the time it can exist within a field.

QUANTUM FIELD

According to quantum physics, particles are excitations of their corresponding fields. For example, electrons are excitations of the electron field.

Particle-antiparticle pairs

The vacuum is not truly nothingness, but frothing with particle-antiparticle pairs that come into existence and then annihilate each other again after an extremely brief period of time.

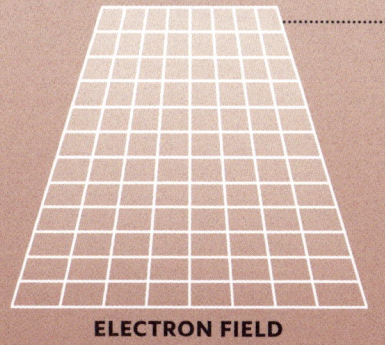

ELECTRON FIELD

TOGETHER BUT APART

Quantum entanglement occurs when two or more particles remain intimately connected to each other, even if separated by vast distances. Despite this separation, a change induced in one will always affect the other, and when the properties of one particle are measured, its partner will instantaneously "know", causing the wave function to collapse (see p.116) as the particle falls to the appropriate corresponding state.

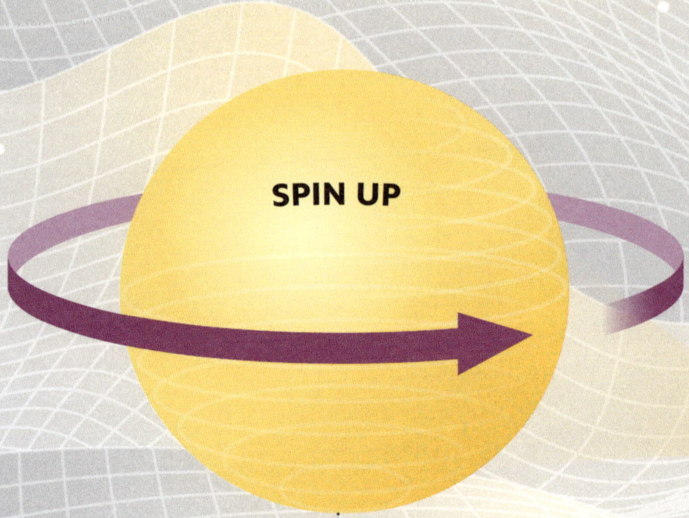

SPIN UP

Correlated particles

A measurement of a physical property – such as spin – of one entangled particle will affect the other. This was described by Albert Einstein as "spooky action at a distance".

PARTICLE A

Measuring the state of particle A will cause the wave function of the entire system to collapse, including, and affecting, particle B. Here, the spin of particle A is measured and determined to be spin up.

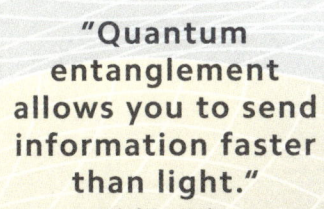

> "Quantum entanglement allows you to send information faster than light."
> Michio Kaku

PARTICLE B

Elsewhere, particle B will now be found in a state that complements particle A. For example, if particle A is in spin up, the correlated spin of particle B will instantaneously be resolved as spin down.

SPIN DOWN

A UNITED FORCE

One of the most ambitious projects in science is the attempt to unify quantum physics, which describes the electromagnetic, weak, and strong forces (see pp.32–33), with general relativity, which describes gravity, into a single model. If this were possible, all the forces in the Universe would be described by a single theory. This theory of everything, as it is known, could describe the four fundamental forces becoming united as a single force at very high energies, such as just after the Big Bang. However, modelling gravity as a quantum force has proved persistently hard, as gravity is its own distinct force, modelled as a curvature in spacetime (see p.136).

GRAND UNIFIED THEORIES

Grand unified theories attempt to merge the strong and electroweak forces into a single force at extremely high energies, corresponding with the very early Universe.

GRAND UNIFIED THEORIES

THEORIES OF EVERYTHING

A theory of everything would describe the force of gravity using quantum physics. It would deal with extreme environments, such as the earliest stages of the Universe.

BIG BANG

QUANTUM GRAVITY

Separating forces

Physicists think that in the early Universe, the four fundamental forces were united. As the Universe expanded and cooled, this single force separated out into the four forces that we encounter today.

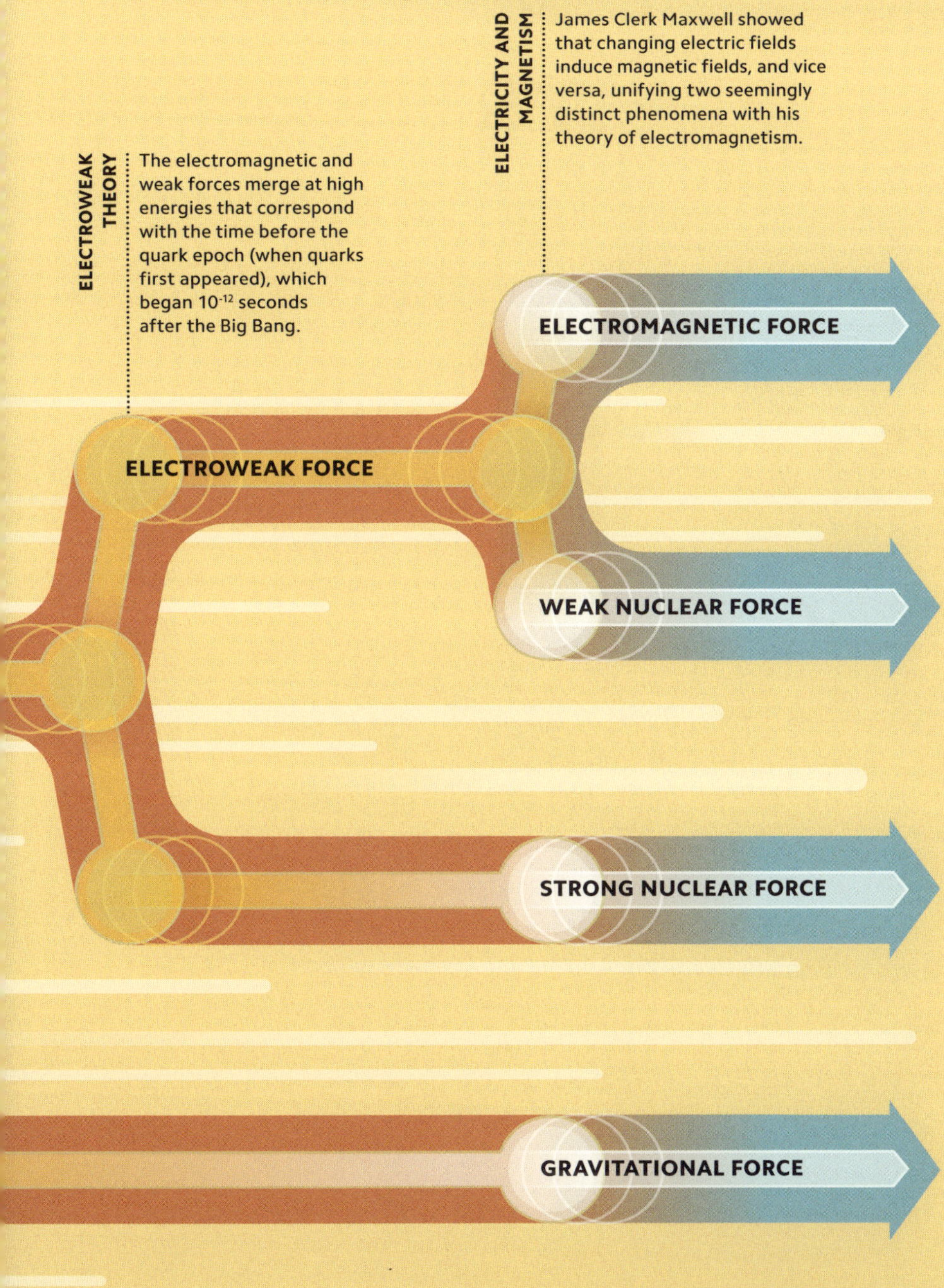

ELECTRICITY AND MAGNETISM
James Clerk Maxwell showed that changing electric fields induce magnetic fields, and vice versa, unifying two seemingly distinct phenomena with his theory of electromagnetism.
ELECTROWEAK THEORY
The electromagnetic and weak forces merge at high energies that correspond with the time before the quark epoch (when quarks first appeared), which began 10^{-12} seconds after the Big Bang.
ELECTROMAGNETIC FORCE
ELECTROWEAK FORCE
WEAK NUCLEAR FORCE
STRONG NUCLEAR FORCE
GRAVITATIONAL FORCE

RELAT

I V I T Y

The special and general theories of relativity describe the laws of physics that apply in some of the most extreme situations. They arise from the fact that light, the ultimate carrier of our information about the Universe, has a finite – though very high – speed that is always the same regardless of how the source moves in relation to the observer. Special relativity describes the effects this creates in situations of relative motion at very high speeds, revealing unexpected flexibility in the dimensions of space and time. General relativity extends these ideas to situations where objects are accelerating, including in gravitational fields.

The simple relativity of classical physics
holds that an observer on a moving ship and
one on dry land would observe phenomena
differently, but agree on simultaneous events.

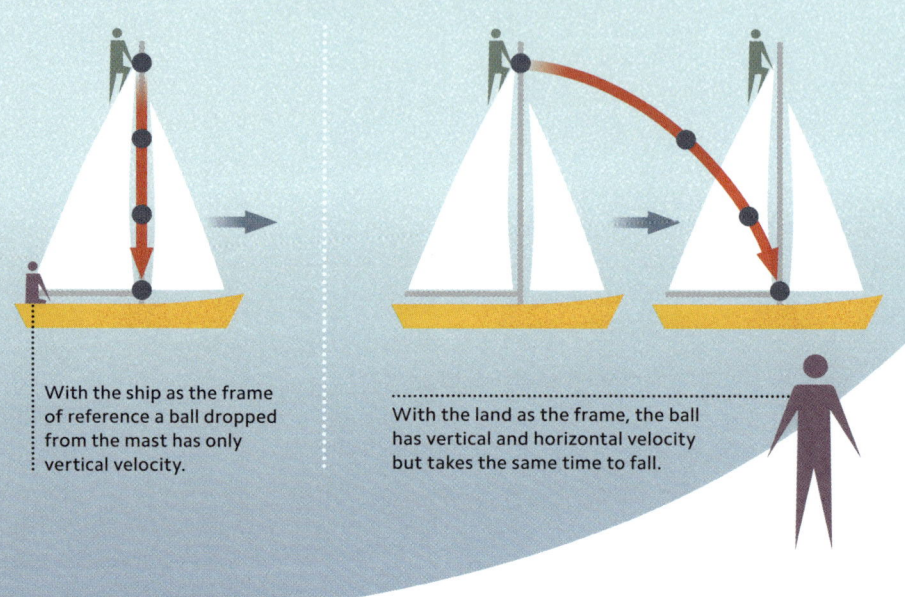

With the ship as the frame
of reference a ball dropped
from the mast has only
vertical velocity.

With the land as the frame, the ball
has vertical and horizontal velocity
but takes the same time to fall.

RELATIVELY SPEAKING

If you are in a closed room, you have no way of knowing if the room is
stationary or moving at a constant speed. Scientists call such a closed
room an inertial frame of reference ("inertial" meaning that the room
is not accelerating). Measurements of velocity depend on which frame
of reference you adopt (see above). In 1905, Albert Einstein proposed
that the speed of light in a vacuum was the same in all frames of
reference – it was fixed regardless of whether the source or observer
was moving. A consequence of his theory is that observers in different
frames of reference may disagree on whether events occur simultaneously.

Special relativity

Einstein's special theory of relativity compares the observation of light from different frames of reference. The fixed speed of light means that the two observers below experience the passage of time in different ways.

OBSERVER INSIDE REFERENCE FRAME

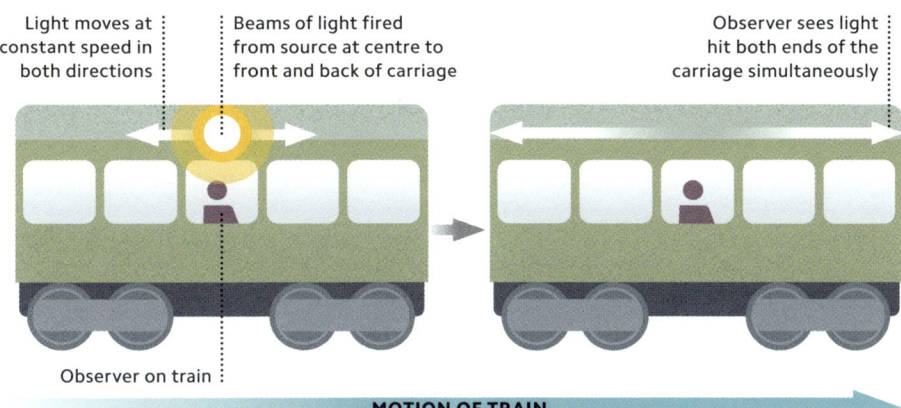

Light moves at constant speed in both directions

Beams of light fired from source at centre to front and back of carriage

Observer sees light hit both ends of the carriage simultaneously

Observer on train

MOTION OF TRAIN

OBSERVER OUTSIDE REFERENCE FRAME

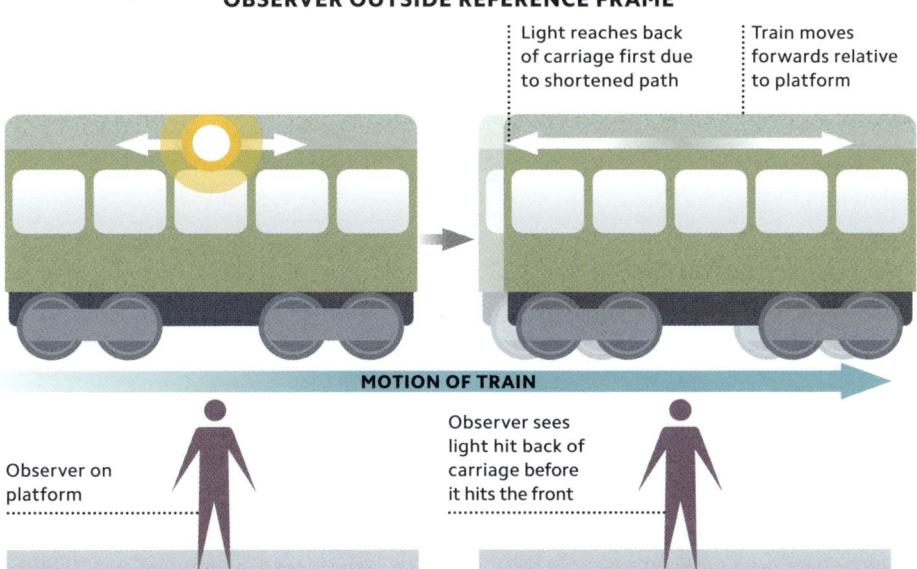

Light reaches back of carriage first due to shortened path

Train moves forwards relative to platform

MOTION OF TRAIN

Observer on platform

Observer sees light hit back of carriage before it hits the front

IN SEARCH OF AETHER

Waves on the surface of a pond exist because the medium – the water – is disturbed. Physicists once thought that a medium that carried light waves would be discovered; they even gave it a name – "luminiferous aether". If the aether exists, however, Earth's motion through it should cause the speed of light from different directions to vary. Clever experiments found no such variations. Einstein's special theory of relativity accepts the speed of light in a vacuum as a constant that is identical for all observers, regardless of their relative motion. This theory has been confirmed experimentally and is a cornerstone of modern science.

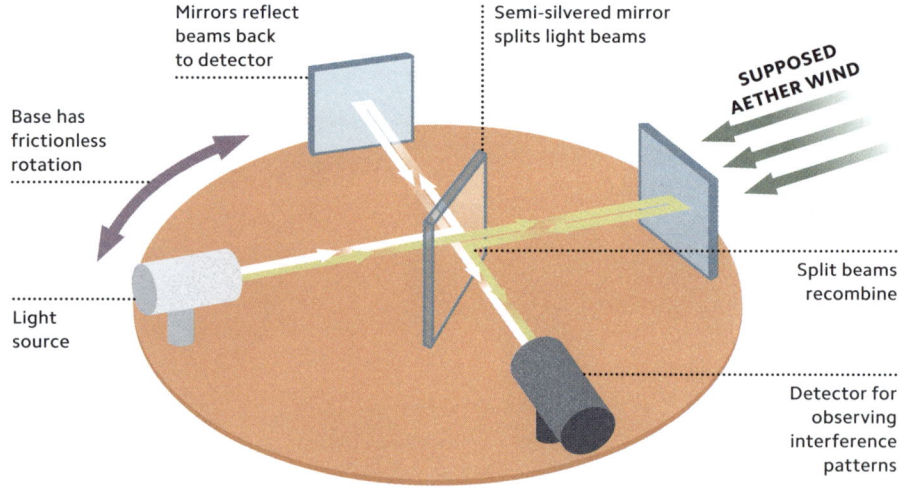

Constant coincidence
Experimenters looking for the aether split a beam of light into two perpendicular paths, then recombined the beam to create interference patterns (see p.115). They expected the beams to arrive at different times depending on their orientation to the flow of aether; however, the beams always arrived simultaneously.

Short measures

Objects moving at relative speeds comparable to that of light appear contracted, but only along the direction of motion.

Rocket cannot contract to zero length because it can never reach the speed of light

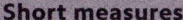

Length of rocket in stationary frame of reference

Passenger on rocket will not experience any contraction in its length whatever its speed relative to Earth

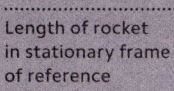

0.9 x SPEED OF LIGHT

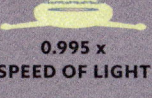

0.995 x SPEED OF LIGHT

0.99995 x SPEED OF LIGHT

Observer in a different frame of reference sees apparent changes in length of the rocket

STATIONARY

MOVING FAST, GETTING SHORTER

Light moving through a vacuum has a fixed speed of 300,000km per second (186,000 miles per second). The special theory of relativity makes a number of predictions for the measurement of objects moving at near-light speeds, including a contraction in length along the direction of motion. Imagine a rocket ship passing Earth at 90 per cent of the speed of light (0.9c). An astronaut on board and an observer on Earth both measure the dimensions of the ship with light whose speed is a constant 1c, but their relative motion means they disagree on a light ray's position at a given moment (see p.129) and therefore on the distance measured.

ABOUT TIME

A consequence of relativity is that observers in different frames of reference disagree on whether events occur simultaneously. This creates an effect called time dilation that becomes obvious for objects moving at close to the speed of light. An observer in one frame of reference sees time running more slowly for objects in another frame moving at near-light speed. For objects moving at 86.6% of the speed of light, time flows at half its "normal" speed.

Time dilation
On a journey at relativistic speeds (close to the speed of light), time passes more slowly for a traveller than for a "fixed" observer.

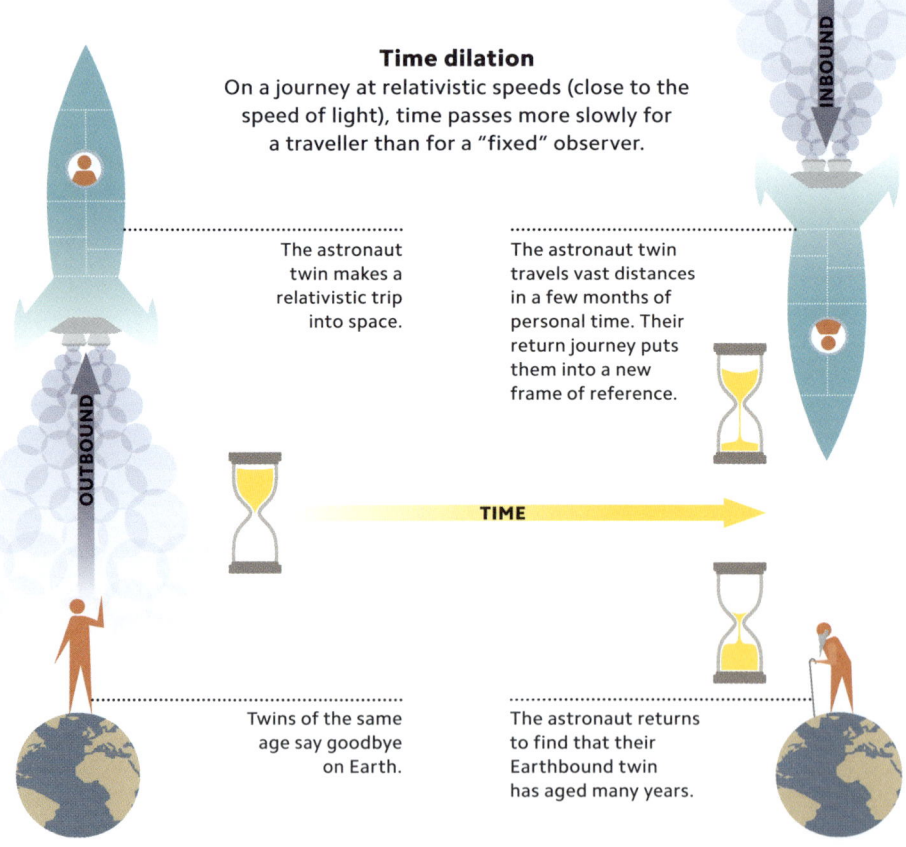

INBOUND

OUTBOUND

The astronaut twin makes a relativistic trip into space.

The astronaut twin travels vast distances in a few months of personal time. Their return journey puts them into a new frame of reference.

TIME

Twins of the same age say goodbye on Earth.

The astronaut returns to find that their Earthbound twin has aged many years.

PAST LIGHT CONE

PAST INFLUENCES
An inverted cone encompasses regions of spacetime whose signals could reach and therefore affect the person in the present.

SPACE

TIME

Regions of influence
A person switches on a light bulb. Taking multiple snapshots of where the light reaches over time gives a series of concentric hoops. Linked together, they form a cone. This cone contains all the possible regions of space and time that the light could interact with in the future.

BEYOND REACH
The person cannot influence areas of spacetime beyond the lightcone, even into the far future, because their signals can never reach them.

FUTURE LIGHT CONE

FUTURE PAST

Classical physics treats time and space as independent dimensions, but relativity sees them as aspects of a four-dimensional structure called spacetime. In spacetime, no object is ever at rest; even those that appear at rest for a certain frame of reference are still moving through the time dimension, and this means they must always maintain a certain resting energy. This energy is given by the famous equation $E = mc^2$.

Gravity and acceleration

The equivalence principle holds that it is impossible to distinguish falling due to gravity from the floor accelerating towards you.

WEIGHTLESS

When the engine switches off and the room stops accelerating, everything within it becomes weightless. A person inside cannot distinguish the effect from freefall on Earth.

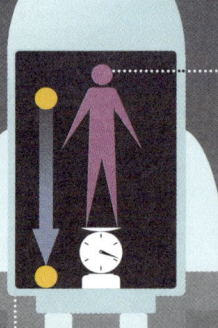

GRAVITY EQUIVALENCE

The person in the accelerating room experiences a downward force equivalent to gravitation.

ON EARTH

A person experiences weight and sees the downward acceleration of objects; both are due to gravity acting on objects with mass.

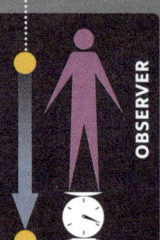

OBSERVER

SEALED ROOM ON EARTH

IN ACCELERATING ROCKET

A person in a room accelerated by a rocket still experiences a force pulling them "downwards" and observes objects accelerating towards the floor.

Earth

"If a person falls freely, he will not feel his own weight."
Albert Einstein

THE GRAVITY OF IT ALL

Einstein's 1915 general theory of relativity expanded on special relativity (see p.129) to consider non-inertial situations – those involving acceleration or deceleration. One of the theory's keystones is the equivalence principle – that an object's gravitational mass (how strongly it interacts with a gravitational field) is equal to its inertial mass. It follows that accelerating frames of reference are equivalent to inertial frames inside gravitational fields. One consequence is that rays of light, despite being massless, bend in the presence of gravity.

Light beam enters window in room

Light paths
The equivalence principle allows the ideas of special relativity to be extended to accelerating frames of reference and to gravitational fields. One important consequence is how light behaves in a gravitational field.

Light beam enters window in room

ACCELERATION
If a ray of light enters the room, an observer within will see the ray follow a curved downward path due to the spacecraft's acceleration.

GRAVITY
Because accelerating reference frames and gravity are equivalent, rays of light must also curve within intense gravitational fields.

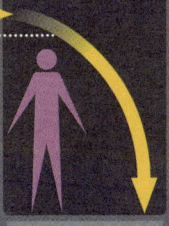

SEALED ROOM ON EARTH

Rocket with very high acceleration

Earth

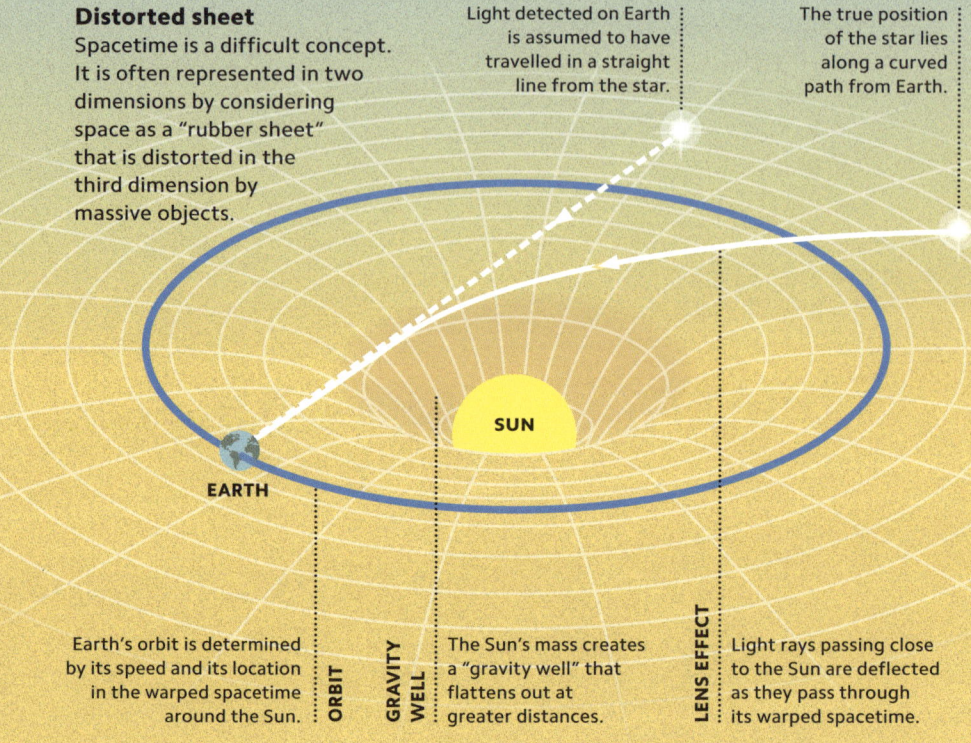

Distorted sheet
Spacetime is a difficult concept. It is often represented in two dimensions by considering space as a "rubber sheet" that is distorted in the third dimension by massive objects.

Light detected on Earth is assumed to have travelled in a straight line from the star.

The true position of the star lies along a curved path from Earth.

SUN

EARTH

Earth's orbit is determined by its speed and its location in the warped spacetime around the Sun.

ORBIT

GRAVITY WELL

The Sun's mass creates a "gravity well" that flattens out at greater distances.

LENS EFFECT

Light rays passing close to the Sun are deflected as they pass through its warped spacetime.

FABRIC OF THE UNIVERSE

In his general theory of relativity, Einstein stated the equivalence of gravitational fields and accelerating frames of reference (see pp.134–35). He showed that space and time are not distinct but form a set of four interdependent dimensions known as spacetime. Where mass and energy are concentrated, spacetime becomes distorted and develops curvature. Massive objects influence their surroundings by their distortion of spacetime, which affects how nearby objects move, rather than through a conventional field of force like electromagnetism.

WRINKLES IN SPACETIME

Massive objects such as stars create symmetrical
distortions of spacetime that flatten out with distance
(see opposite). However, general relativity also predicts another
type of distortion – gravitational waves. These are tiny shifts in the
dimensions of space that spread out at the speed of light. They are
generated only in extreme situations – for example, when supernovas
explode or black holes orbit one other in tight pairs. They can
be detected by instruments called interferometers.

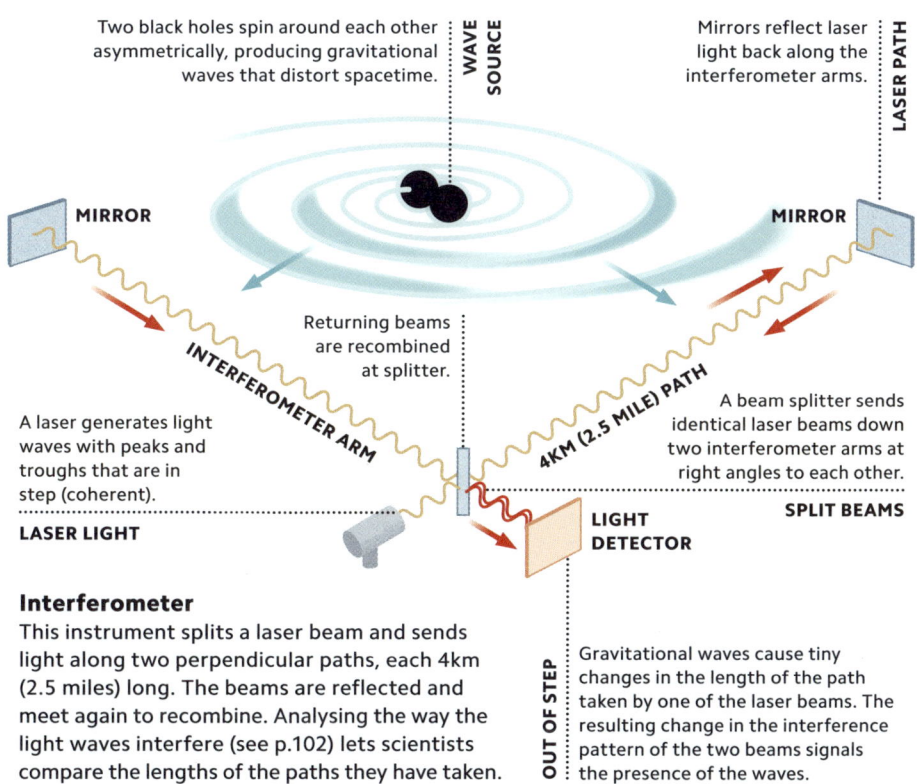

Two black holes spin around each other
asymmetrically, producing gravitational
waves that distort spacetime.

WAVE SOURCE

Mirrors reflect laser
light back along the
interferometer arms.

LASER PATH

MIRROR

MIRROR

Returning beams
are recombined
at splitter.

INTERFEROMETER ARM

4KM (2.5 MILE) PATH

A laser generates light
waves with peaks and
troughs that are in
step (coherent).

A beam splitter sends
identical laser beams down
two interferometer arms at
right angles to each other.

LASER LIGHT

SPLIT BEAMS

LIGHT DETECTOR

OUT OF STEP

Interferometer

This instrument splits a laser beam and sends
light along two perpendicular paths, each 4km
(2.5 miles) long. The beams are reflected and
meet again to recombine. Analysing the way the
light waves interfere (see p.102) lets scientists
compare the lengths of the paths they have taken.

Gravitational waves cause tiny
changes in the length of the path
taken by one of the laser beams. The
resulting change in the interference
pattern of the two beams signals
the presence of the waves.

ASTROP
AND
COSMOL

HYSICS

OGY

The laws of physics that we experience on Earth extend across the Universe, allowing us to understand processes and objects that we cannot hope to probe at close quarters. These include the birth, evolution, and death of stars of various kinds, and the strange stellar remnants they leave behind. Armed with the ability to predict the behaviour and properties of stars and even entire galaxies, astronomers can also discover patterns in the Universe on the largest scales. Ultimately, the specialized field of cosmology can allow us to understand the origin, current state, and possible fates of our entire Universe.

Views of the cosmos

On large scales, the Universe appears uniform in every direction to an observer on Earth. Zooming in to areas of the cosmos (see right) reveals ever more levels of structure, where galaxies and clusters of galaxies are held together by gravity.

UNIFORM UNIVERSE

Earth does not occupy a privileged spot in the Universe. The cosmological principle holds that all locations in the Universe are much the same and that matter is distributed uniformly (when viewed on a large enough scale). Furthermore, the Universe exhibits similar properties in every direction. The principle is important to astronomers because it means that studies made from Earth should be applicable to the cosmos as a whole.

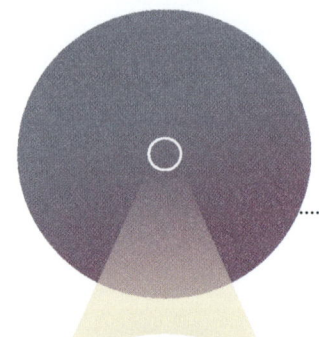

The finite speed of light limits our view to objects less than 13.8 billion light years away. At this scale, the Universe appears uniform.

OUR OBSERVABLE UNIVERSE

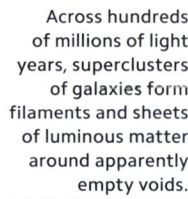

Across hundreds of millions of light years, superclusters of galaxies form filaments and sheets of luminous matter around apparently empty voids.

FILAMENTS AND VOIDS

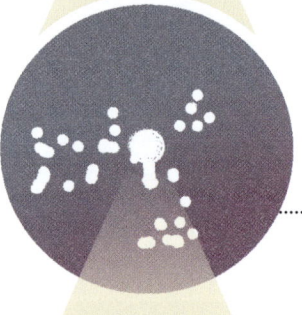

We are located in a galaxy supercluster some 100 million light-years across. Superclusters are, in turn, made up of smaller groups of galaxies called galaxy clusters.

SUPERCLUSTERS

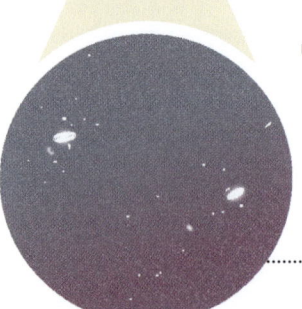

In our region of the Universe, most matter is concentrated in three large galaxies and several smaller ones, forming a loose galaxy cluster called the Local Group; this is some 10 million light-years across.

LOCAL GROUP

YOUNG STARS

Stars are enormous balls of hot gas. The extreme temperatures and pressures in their cores drive nuclear fusion of lighter elements into heavier ones. This releases energy, which heats the core and produces radiation that escapes outwards through the star; these processes counteract gravitational forces and so prevent the star from collapsing in on itself. Star formation begins in vast but sparse interstellar clouds of gas and dust, dominated by the lightweight elements hydrogen and helium. Within these clouds, where matter is more concentrated, its gravitational pull is sufficient to cause it to collapse inwards.

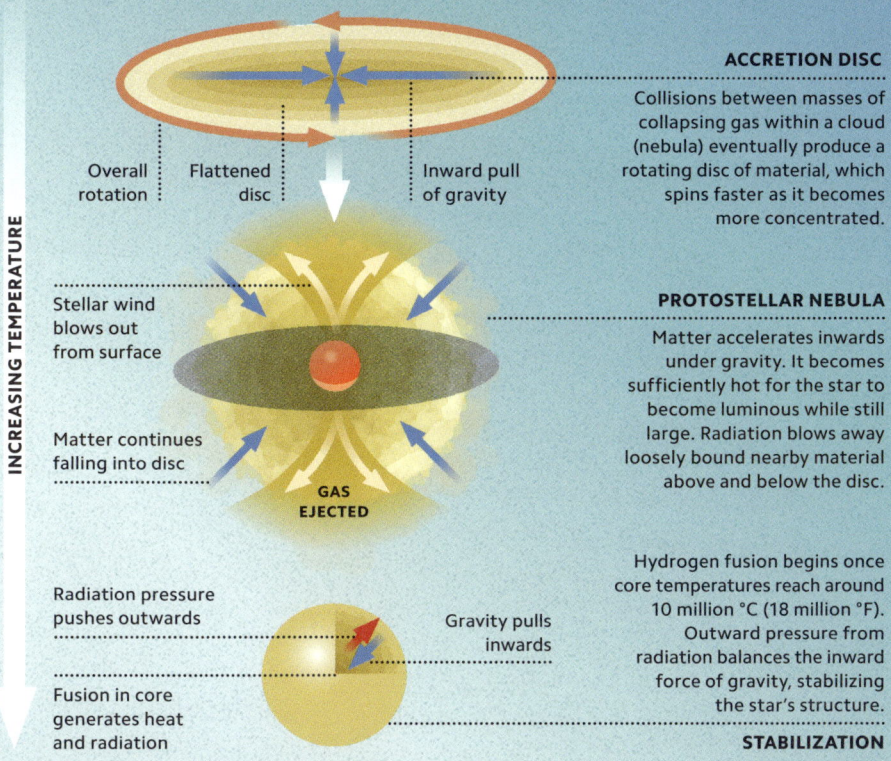

INCREASING TEMPERATURE

ACCRETION DISC

Overall rotation | Flattened disc | Inward pull of gravity

Collisions between masses of collapsing gas within a cloud (nebula) eventually produce a rotating disc of material, which spins faster as it becomes more concentrated.

Stellar wind blows out from surface

PROTOSTELLAR NEBULA

Matter accelerates inwards under gravity. It becomes sufficiently hot for the star to become luminous while still large. Radiation blows away loosely bound nearby material above and below the disc.

Matter continues falling into disc

GAS EJECTED

Radiation pressure pushes outwards

Gravity pulls inwards

Hydrogen fusion begins once core temperatures reach around 10 million °C (18 million °F). Outward pressure from radiation balances the inward force of gravity, stabilizing the star's structure.

Fusion in core generates heat and radiation

STABILIZATION

SHINING BRIGHT

Stars vary greatly in brightness, colour, and size. Some of these differences result from how the stars formed, while others come about as stars enter different phases of their lives. A famous chart called the Hertzsprung–Russell diagram compares the luminosity (total energy output) of stars with their surface temperature (or colour). The most prominent feature of the chart is a diagonal band of stars – from brilliant hot blue stars to faint, cool red ones. A star spends most of its lifetime somewhere on this main sequence band. During this time it is stable, shining by the fusion of hydrogen to helium.

Hertzsprung-Russell diagram
The amount of power produced by a star (in terms of light output) is related to the temperature of its surface and its area. Red stars are cooler than blue ones; so if a red and blue star have a similar luminosity (energy output), the red star must have a larger surface area than the blue.

BRIGHTER

BLUE SUPERGIANT

HOT BLUE STARS
Stars somewhat heavier than the Sun fuse hydrogen rapidly; they are hotter and more luminous than the Sun, but burn through their fuel in tens or hundreds of millions of years.

BRIGHTNESS

THE SUN

DIMMER

MAIN SEQUENCE
Stars spend most of their lives on the main sequence. The rate of nuclear fusion at a star's core largely depends on its mass.

HOTTER

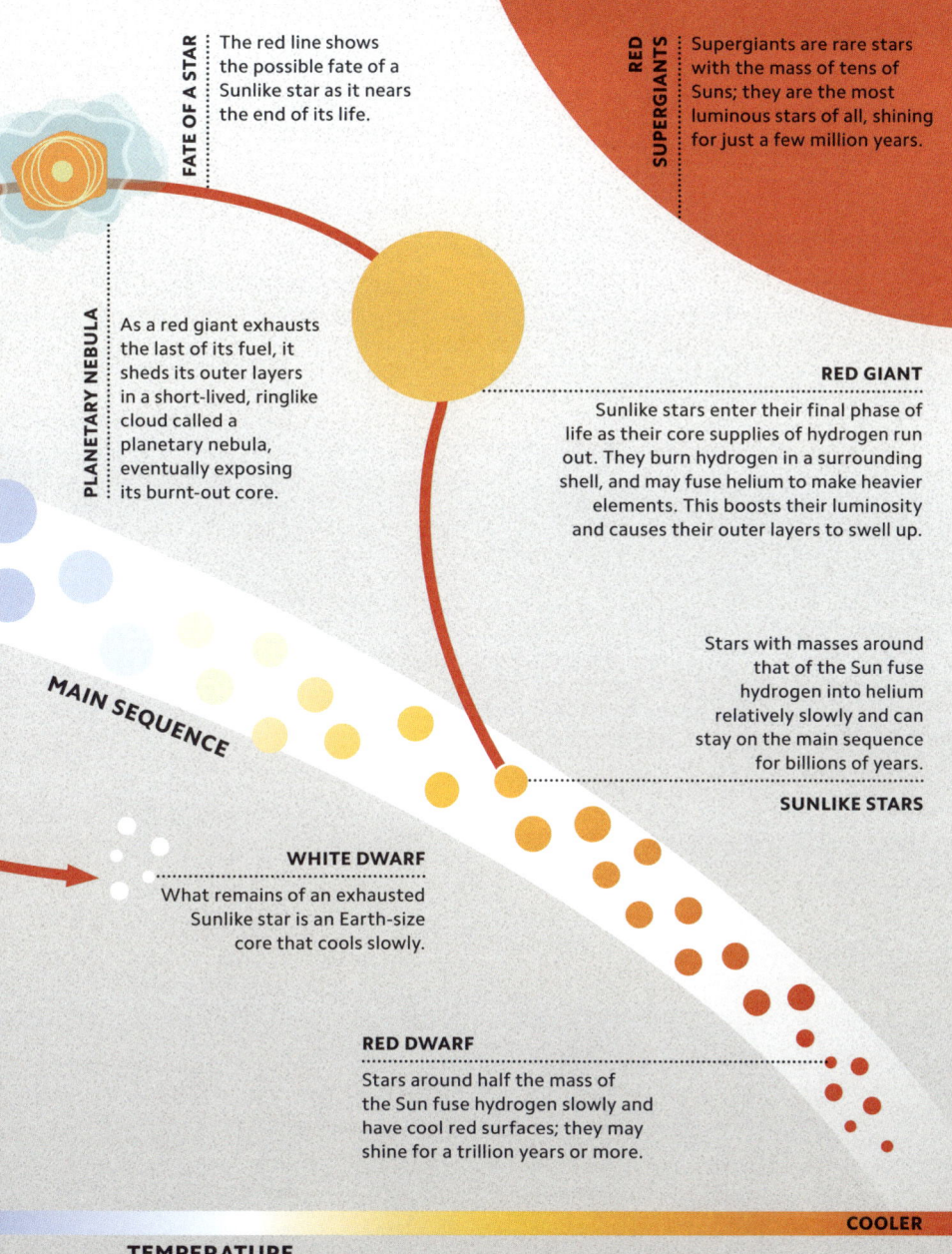

The red line shows the possible fate of a Sunlike star as it nears the end of its life.

RED SUPERGIANTS

Supergiants are rare stars with the mass of tens of Suns; they are the most luminous stars of all, shining for just a few million years.

PLANETARY NEBULA

As a red giant exhausts the last of its fuel, it sheds its outer layers in a short-lived, ringlike cloud called a planetary nebula, eventually exposing its burnt-out core.

RED GIANT

Sunlike stars enter their final phase of life as their core supplies of hydrogen run out. They burn hydrogen in a surrounding shell, and may fuse helium to make heavier elements. This boosts their luminosity and causes their outer layers to swell up.

Stars with masses around that of the Sun fuse hydrogen into helium relatively slowly and can stay on the main sequence for billions of years.

MAIN SEQUENCE

SUNLIKE STARS

WHITE DWARF

What remains of an exhausted Sunlike star is an Earth-size core that cools slowly.

RED DWARF

Stars around half the mass of the Sun fuse hydrogen slowly and have cool red surfaces; they may shine for a trillion years or more.

COOLER

TEMPERATURE

STAR SIZE, TEMPERATURE, AND BRIGHTNESS | 143

ORBITAL DANCE

Some stars are bound together by gravity from the time they form (see p.141). In a binary system, two stars are locked in a gravitational dance around one another, following elliptical orbits around a shared "centre of mass". Their double nature can be detected by observing Doppler shifts in the colour of their light as they move around their orbits.

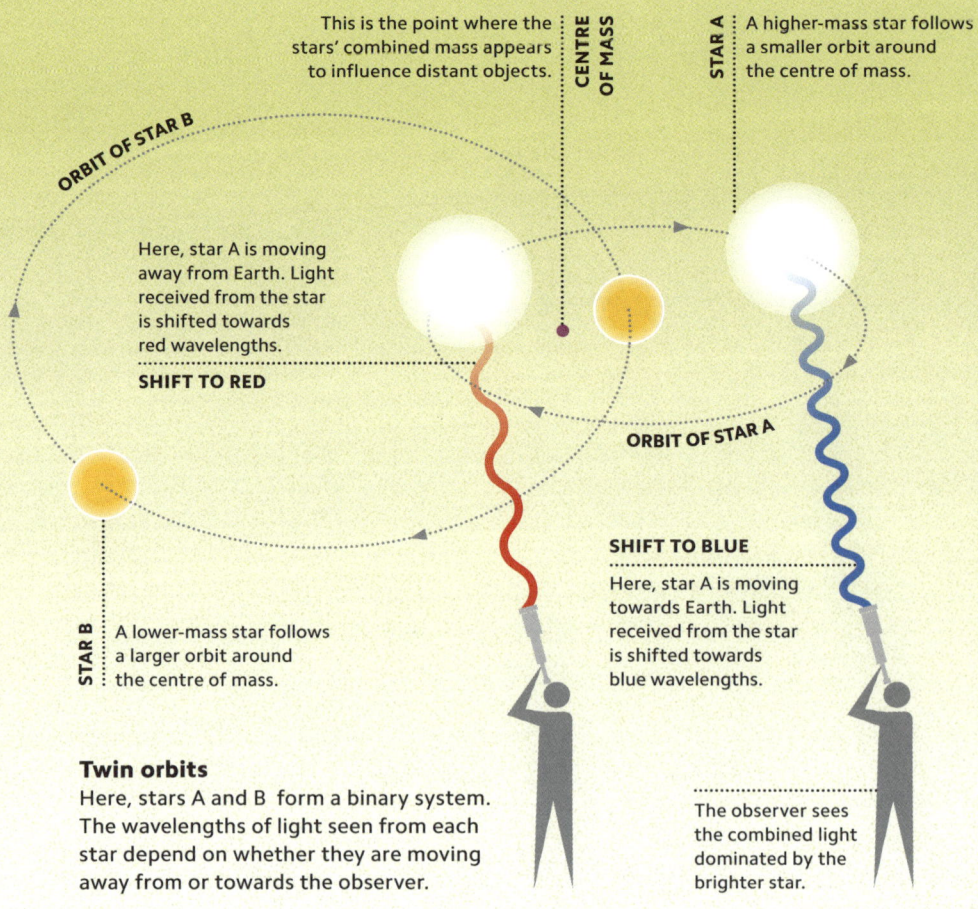

CENTRE OF MASS
This is the point where the stars' combined mass appears to influence distant objects.

STAR A
A higher-mass star follows a smaller orbit around the centre of mass.

ORBIT OF STAR B

Here, star A is moving away from Earth. Light received from the star is shifted towards red wavelengths.

SHIFT TO RED

ORBIT OF STAR A

STAR B
A lower-mass star follows a larger orbit around the centre of mass.

SHIFT TO BLUE
Here, star A is moving towards Earth. Light received from the star is shifted towards blue wavelengths.

Twin orbits
Here, stars A and B form a binary system. The wavelengths of light seen from each star depend on whether they are moving away from or towards the observer.

The observer sees the combined light dominated by the brighter star.

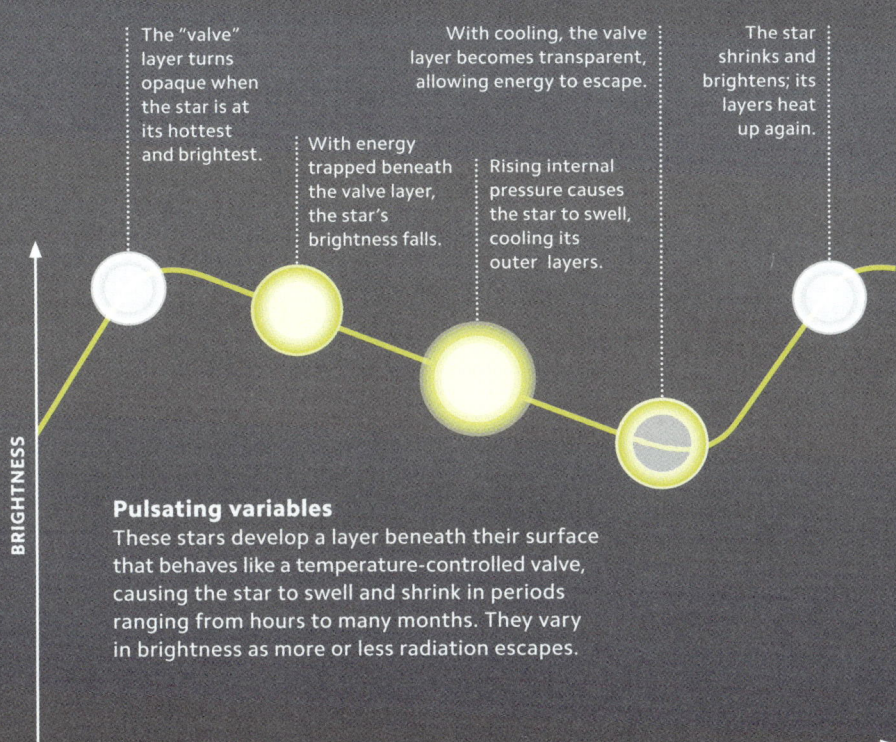

The "valve" layer turns opaque when the star is at its hottest and brightest.

With cooling, the valve layer becomes transparent, allowing energy to escape.

The star shrinks and brightens; its layers heat up again.

With energy trapped beneath the valve layer, the star's brightness falls.

Rising internal pressure causes the star to swell, cooling its outer layers.

BRIGHTNESS

Pulsating variables
These stars develop a layer beneath their surface that behaves like a temperature-controlled valve, causing the star to swell and shrink in periods ranging from hours to many months. They vary in brightness as more or less radiation escapes.

TIME

LIGHT CYCLES

Although the nuclear reactions that power stars change very slowly, stars may still vary in brightness on much shorter timescales. Binary stars (see opposite) can eclipse one other, causing their combined light output to fluctuate. Other stars have powerful magnetic fields that release vast amounts of energy in brilliant flares, or show dark or bright patches on their surfaces as they rotate. Many more stars, however, pass through a pulsating phase where their upper layers become unstable, driving cyclical changes in size and brightness.

Low-mass stars

Stars with a mass of between 0.8 and eight Suns may live for billions of years before becoming red giants.

SLOW BURN

Low-mass stars tend to be reddish in colour and cooler than larger stars.

Hydrogen fusion moves from the core to a shell around the star; helium fusion may ignite in the core. The star swells to great size and becomes unstable.

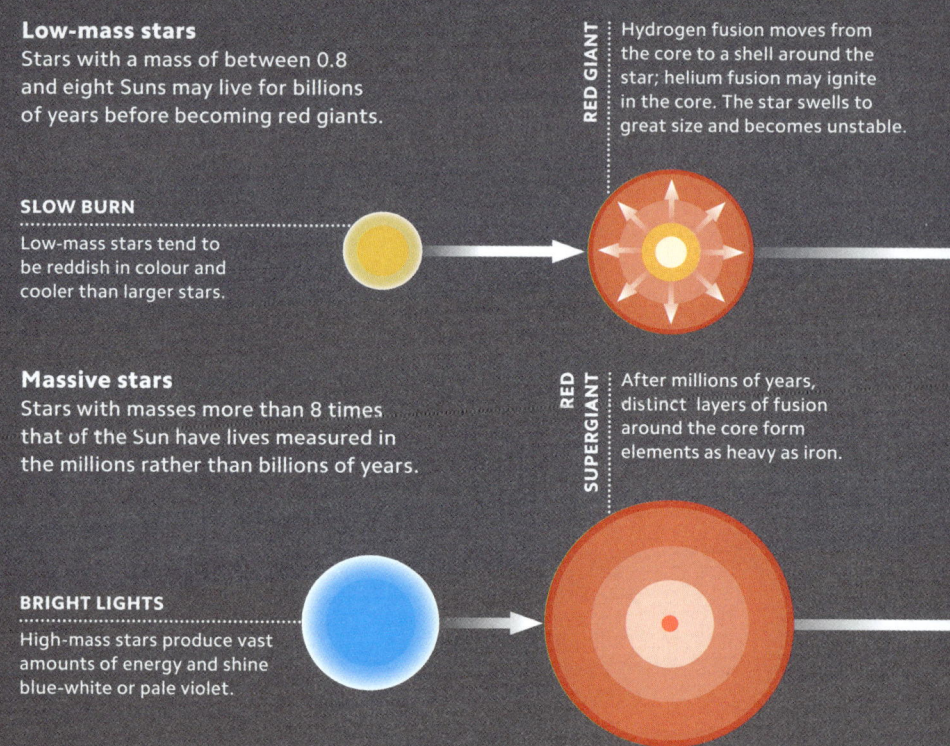

Massive stars

Stars with masses more than 8 times that of the Sun have lives measured in the millions rather than billions of years.

After millions of years, distinct layers of fusion around the core form elements as heavy as iron.

BRIGHT LIGHTS

High-mass stars produce vast amounts of energy and shine blue-white or pale violet.

STELLAR FATES

Stars have finite lives. As a star's core runs out of hydrogen fuel, it collapses and so becomes hotter, allowing fusion of hydrogen outside the core. The star continues to shine. A star of similar size to the Sun eventually becomes unstable, shedding its outer layers as its exhausted core slowly collapses to become a compact white dwarf. A more massive star continues fusion for longer, building a layered core of heavier elements. When fusion can no longer continue, the core collapses abruptly; shockwaves produce a powerful supernova explosion and only a tiny, superdense object – a neutron star or black hole – survives.

PLANETARY NEBULA
A planetary nebula (cloud) forms as the star sheds its outer layers into space and its hot core is exposed.

WHITE DWARF
A white dwarf is formed. This is the burnt-out but still hot remnant of the star's core.

BLACK DWARF
Over billions of years, a white dwarf cools to become a black dwarf.

SUPERNOVA
Exhausted of fuel, the core collapses, producing a shockwave that heats and compresses the outer layers.

Jets of matter escape from the disc around the black hole.

Singularity

The gravitational pull of a black hole prevents the escape even of photons.

BLACK HOLE
An exceptionally large star may collapse into a super-dense black hole; the core dwindles to a tiny point called a singularity.

NEUTRON STAR
Supernovas usually give rise to ultra-dense neutron stars, so called because their electrons and protons combine into neutrons.

Intense magnetism around some neutron stars causes the emission of beams of radiation. Such stars are called pulsars.

A teaspoon of matter from a neutron star would weigh 10 million tons.

MEASURING THE UNIVERSE

The closest star to the Sun is Proxima Centauri, 2.4 light years away, while the furthest observable objects are galaxies about 13.4 billion light years distant. Astronomers use a variety of techniques to measure such vast spans. Parallax measurement (see below) gives the distances to relatively nearby stars. The distance to more remote objects can be found by using reference objects known as standard candles that have a known absolute (or inherent) luminosity (see right). A standard candle's apparent brightness reveals its distance from Earth. Measuring redshift (see p.150) provides another way to assess the distance of the furthest visible galaxies.

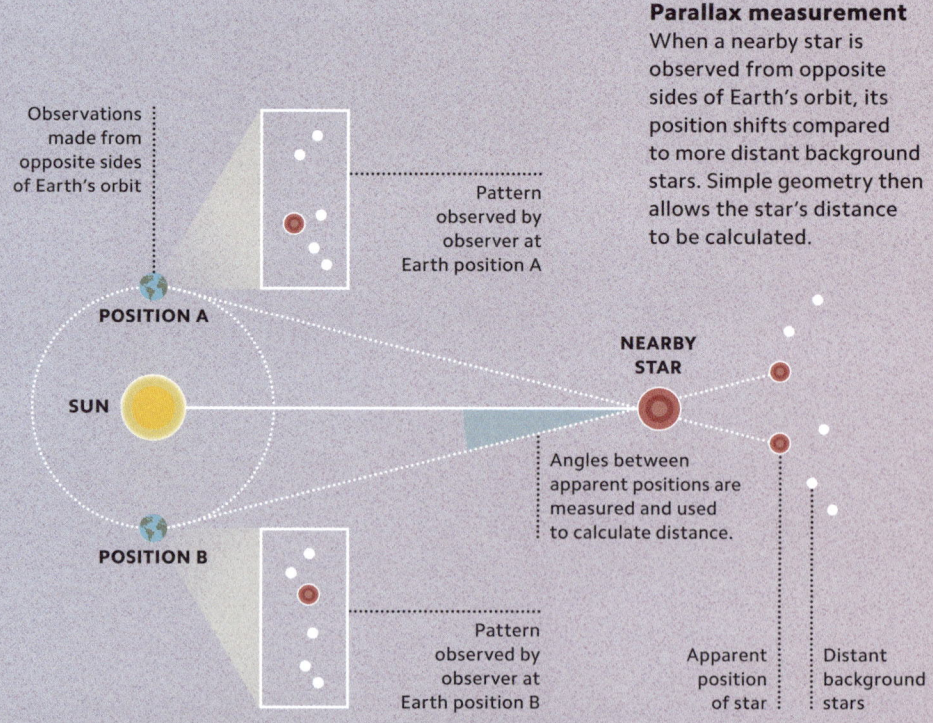

Observations made from opposite sides of Earth's orbit

Pattern observed by observer at Earth position A

POSITION A

SUN

POSITION B

Pattern observed by observer at Earth position B

Parallax measurement
When a nearby star is observed from opposite sides of Earth's orbit, its position shifts compared to more distant background stars. Simple geometry then allows the star's distance to be calculated.

NEARBY STAR

Angles between apparent positions are measured and used to calculate distance.

Apparent position of star

Distant background stars

COUNTING PHOTONS

Astronomical instruments count the number of photons passing through a standard area.

RADIATION INTENSITY

1

16 PHOTONS

DISTANCE FROM SOURCE

Light source of known absolute luminosity

The number of photons passing through a standard area falls with the square of the distance travelled.

$\frac{1}{4}$

RADIATION INTENSITY

1

INVERSE SQUARE

As distance increases by a factor of two, photons are spread over an area four times greater, so the apparent luminosity drops by a factor of four.

FALLING LUMINOSITY

4 PHOTONS

DISTANCE FROM SOURCE

$\frac{1}{16}$

RADIATION INTENSITY

$\frac{1}{9}$

$\frac{1}{4}$

1

1 PHOTON

DISTANCE FROM SOURCE

DISTANCE

Applying the inverse square law to the number of photons detected in a standard area allows astronomers to calculate the distance to the source.

Standard candle measurement

A standard candle has a known luminosity or light output. Because its radiation gets more thinly spread at greater distances, its apparent brightness to an observer on Earth reveals its distance.

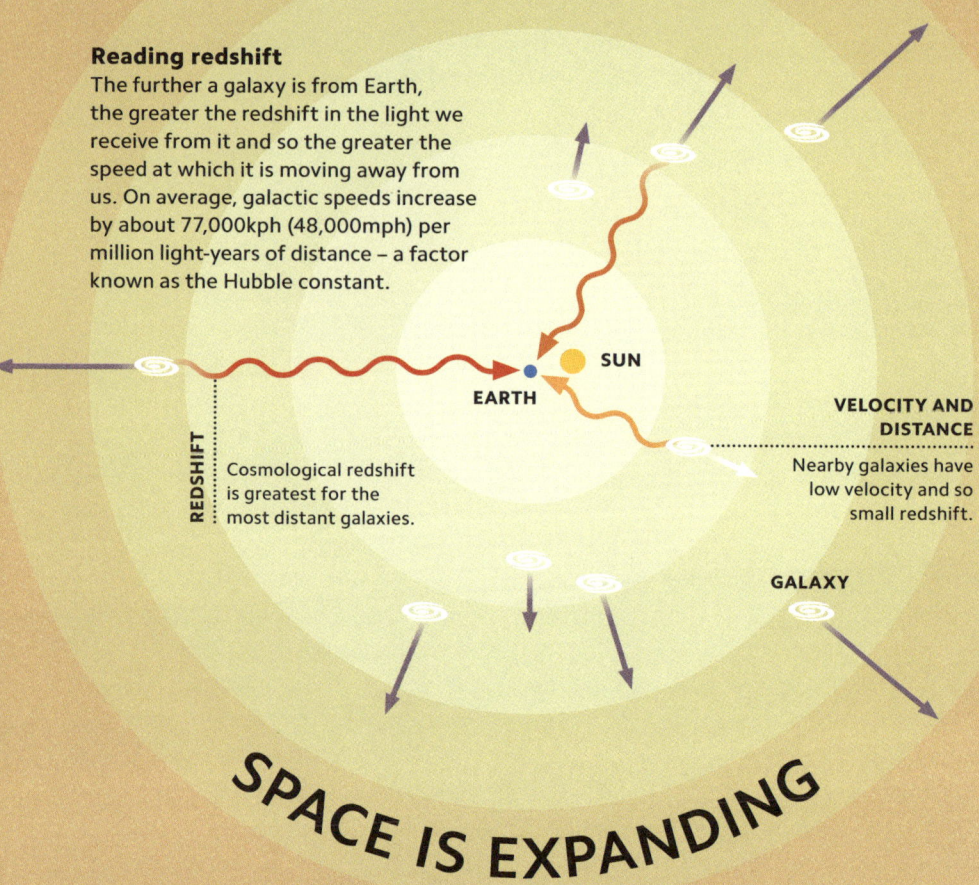

Reading redshift
The further a galaxy is from Earth, the greater the redshift in the light we receive from it and so the greater the speed at which it is moving away from us. On average, galactic speeds increase by about 77,000kph (48,000mph) per million light-years of distance – a factor known as the Hubble constant.

SUN

EARTH

REDSHIFT

Cosmological redshift is greatest for the most distant galaxies.

VELOCITY AND DISTANCE

Nearby galaxies have low velocity and so small redshift.

GALAXY

SPACE IS EXPANDING

By measuring the Doppler shift in the light from distant galaxies (see p.144), astronomers have deduced their speeds. Such studies have revealed a remarkable pattern – that, aside from the very nearest, all galaxies are moving away from Earth, and their speed of retreat is proportional to their distance from Earth. This relationship, known as Hubble's law, is evidence that the Universe as a whole is expanding. Galaxies are not moving away from Earth across space; instead, space itself is expanding and carrying galaxies with it.

ALL CLEAR
··
Light rays were able to move
freely only 380,000 years after
the Big Bang (see p.152).

THE START OF ALL THINGS

According to current understanding, the Universe originated some 13.8 billion years ago from a very dense, hot, and uniform state. All matter and energy, along with time and space themselves, began with a violent event called the Big Bang. For a few minutes, extreme temperatures and energies allowed the creation of strange particles, which flashed in and out of existence. As space expanded and temperatures fell, these particles bound together into protons, neutrons, atomic nuclei, and eventually atoms themselves. The Universe is still expanding, but much of its matter has been drawn together by gravity to form galaxies, stars, and planets.

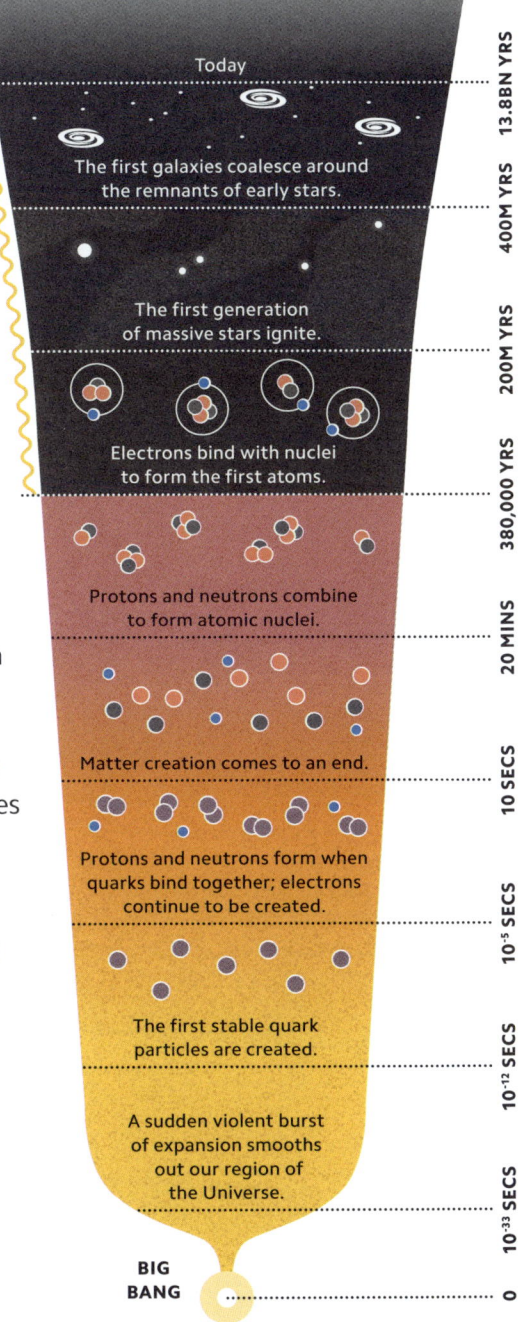

Today

The first galaxies coalesce around the remnants of early stars.

The first generation of massive stars ignite.

Electrons bind with nuclei to form the first atoms.

Protons and neutrons combine to form atomic nuclei.

Matter creation comes to an end.

Protons and neutrons form when quarks bind together; electrons continue to be created.

The first stable quark particles are created.

A sudden violent burst of expansion smooths out our region of the Universe.

BIG BANG

13.8BN YRS

400M YRS

200M YRS

380,000 YRS

20 MINS

10 SECS

10^{-5} SECS

10^{-12} SECS

10^{-33} SECS

0

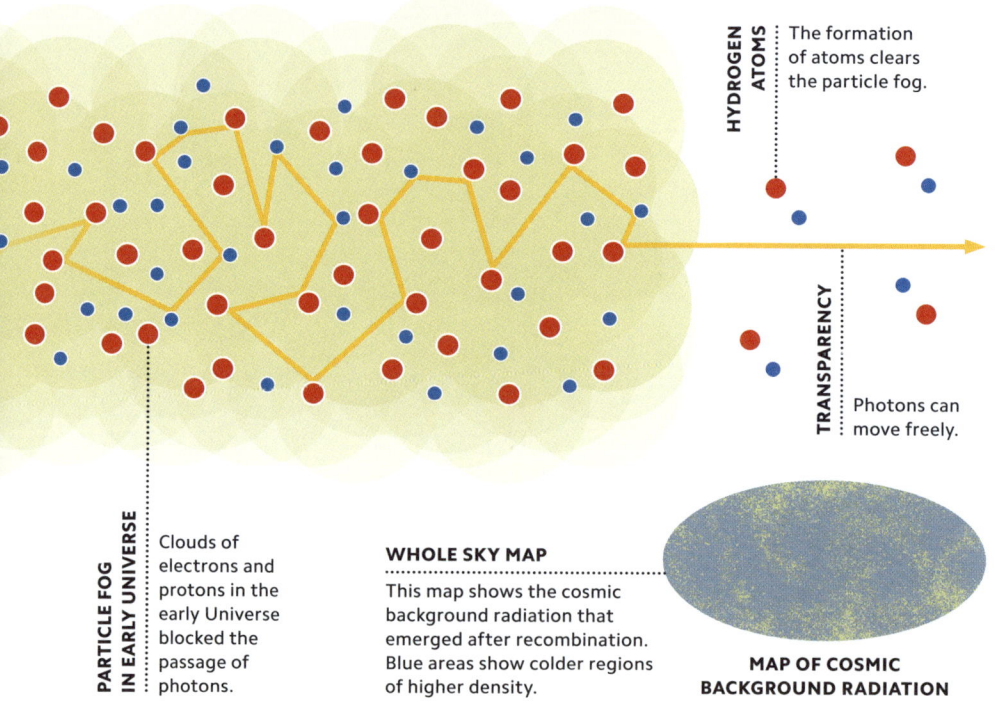

HYDROGEN ATOMS

The formation of atoms clears the particle fog.

TRANSPARENCY

Photons can move freely.

PARTICLE FOG IN EARLY UNIVERSE

Clouds of electrons and protons in the early Universe blocked the passage of photons.

WHOLE SKY MAP

This map shows the cosmic background radiation that emerged after recombination. Blue areas show colder regions of higher density.

MAP OF COSMIC BACKGROUND RADIATION

FIRST MAP OF THE UNIVERSE

Some 380,000 years after the Big Bang (see p.151), the Universe underwent a key transition known as recombination. Before this time, photons – quanta of light – were trapped within a dense fog of protons and electrons. In recombination, these particles bound together, forming the first atoms. This caused a reduction in particle density, which allowed the passage of photons. The Universe became transparent. Radiation that escaped during recombination still permeates the Universe today, but its wavelength has been redshifted (see p.150) into the microwave part of the spectrum. Scientists map its fluctuations to study density variations in the early Universe.

THE HIDDEN UNIVERSE

About 84 per cent of all the mass in the Universe is invisible "dark matter". It is thought to consist of particles that do not interact with the electromagnetic force (see pp.32–33). Recent studies of the distant Universe also reveal the existence of so-called "dark energy" – a mysterious effect that is driving the continued and accelerating expansion of the Universe (see p.150).

Dark matter

Visible matter

Dark energy

Composition of the Universe
Dark matter outweighs normal matter by about six to one. The remainder of the Universe is made up of dark energy.

Dark matter revealed
Studies of the velocities at which stars orbit the centre of a spiral galaxy suggest the presence of invisible matter.

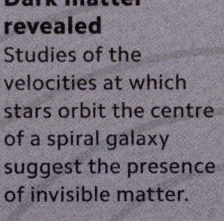

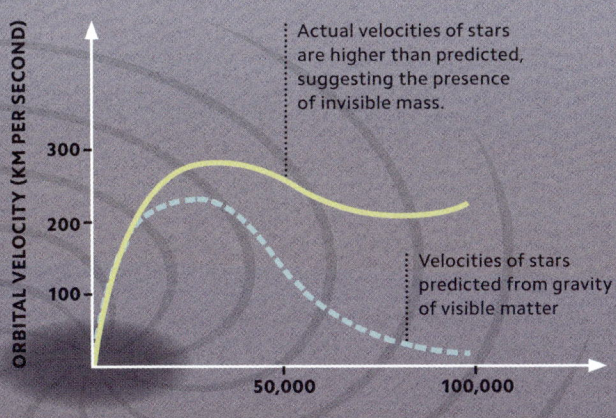

ORBITAL VELOCITY (KM PER SECOND)

Actual velocities of stars are higher than predicted, suggesting the presence of invisible mass.

300

200

100

Velocities of stars predicted from gravity of visible matter

50,000   100,000

DISTANCE FROM CENTRE OF GALAXY (LIGHT YEARS)

SPIRAL GALAXY

MATTER
DISINTEGRATES

Dark energy eventually overcomes electromagnetic forces that hold atoms and molecules together.

Expansion will overwhelm local gravity on ever smaller scales.

CONTINUED EXPANSION

RIP, CHILL, OR CRUNCH

Many billions of years from now, the ultimate fate of the Universe may depend on several factors: the expansion initiated by the Big Bang; the mass and gravity of all the matter it contains; and the behaviour of the dark energy (see p.153) that is currently driving an increasing expansion of space. Depending on the relative strength of these factors, the Universe may collapse back on itself, tear itself apart, or simply continue expanding forever until the fuel for new stars is exhausted and matter itself slowly disintegrates through quantum decay.

PRESENT

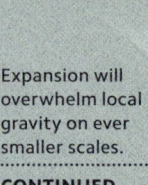

BIG BANG

Big Rip
Dark energy seems to be increasing in strength. If this is so, the expansion of space may overwhelm the forces that bind matter together.

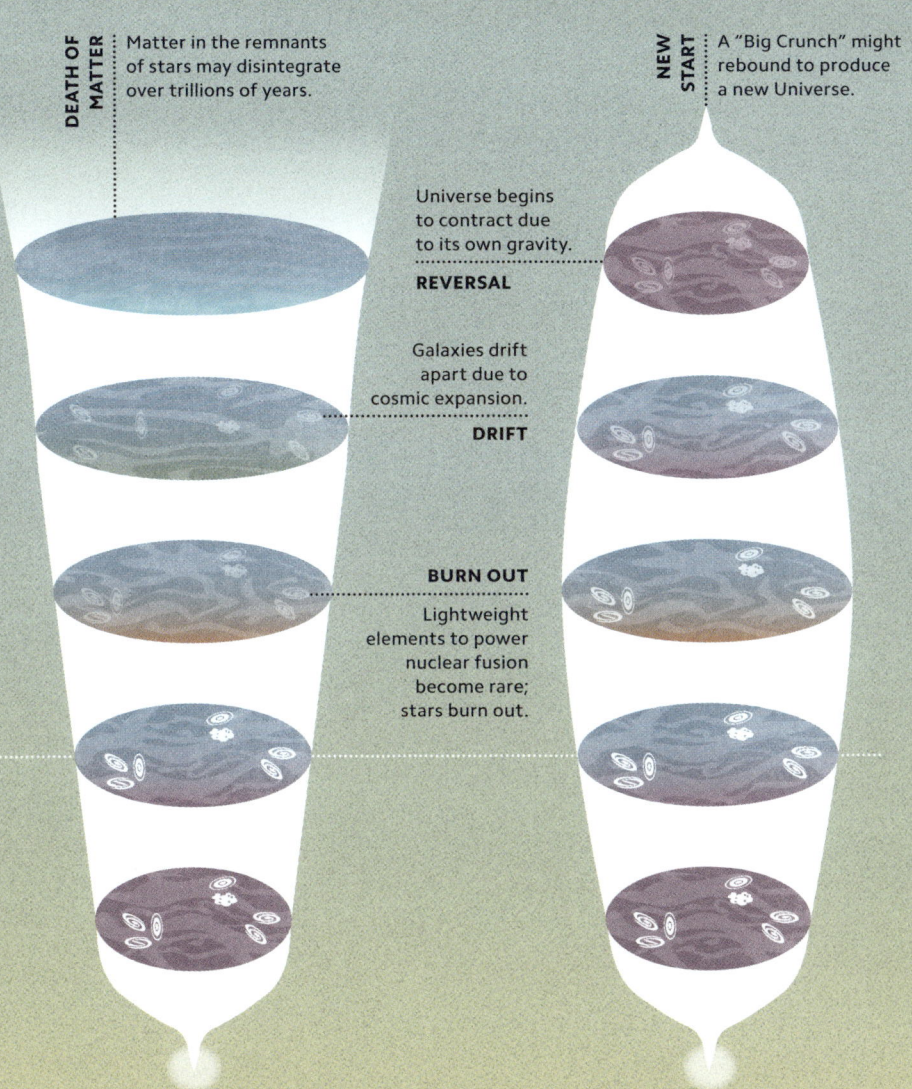

DEATH OF MATTER Matter in the remnants of stars may disintegrate over trillions of years.

NEW START A "Big Crunch" might rebound to produce a new Universe.

Universe begins to contract due to its own gravity.

REVERSAL

Galaxies drift apart due to cosmic expansion.

DRIFT

BURN OUT

Lightweight elements to power nuclear fusion become rare; stars burn out.

Big Chill
The Universe continues to expand but gradually "runs down" to a state of zero thermodynamic free energy.

Big Crunch
If the Universe contains sufficient mass, and dark energy fades or reverses, cosmic expansion may reverse, drawing everything back to a hot, dense, high-energy state.

INDEX

Page numbers in **bold** refer to main entries.

ACKNOWLEDGMENTS

DK would like to thank the following for their help with this book: Noor Ali for design assistance; Katie John for proofreading; and Vanessa Bird for the index. DTP Designer: Raman Panwar. Senior DTP Designer: Harish Aggarwal. Senior jackets coordinator: Priyanka Sharma Saddi.

Cover images: front and back:
Getty Images/iStock:
Veronika Oliinyk, SpicyTruffel

SIMPLY EXPLAINED

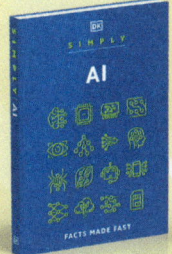

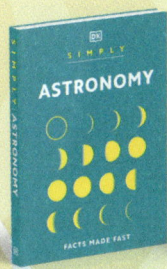

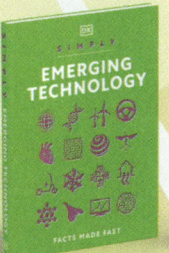

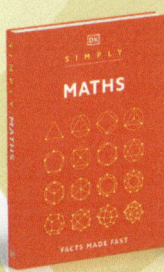

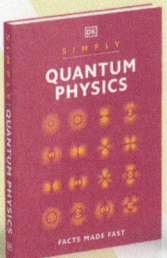